CASES IN BIOCHEMISTRY

Kathleen Cornely

Providence College

JOHN WILEY & SONS, INC.

New York • Chichester • Weinheim • Brisbane • Singapore • Toronto

The front cover shows some of the molecular assemblies that form the circle of life:
DNA makes RNA makes protein makes DNA.

The images are (clockwise from the top):
1. B-DNA, *based on an X-ray structure by Richard Dickerson and Horace Drew.*
2. The nucleosome, *courtesy of Timothy Richmond.*
3. Model of the *lac* repressor in complex with DNA and CAP protein, *courtesy of Ponzy Lu and Mitchell Lewis.*
4. Ribozyme RNA, *based on an X-ray structure by Jennifer Doudna.*
5. The ribosome in complex with tRNAs, *courtesy of Joachim Frank.*
6. DNA polymerase in complex with DNA, *courtesy of Tom Ellenberger.*

The central image is based on Leonardo da Vinci's drawing *Study of Proportions.*
It represents for us the never ending human quest for understanding. (© G. Bartholomew/Westlight)

ISBN 0-471-32283-0

Printed in the United States of America

10 9 8 7 6 5 4 3 2 1

Printed and bound by Hamilton Printing Company

Preface

The case study method of instruction has been increasing in popularity in the last several years. In a recent article in the *Journal of College Science Teaching*, Clyde Freeman Herreid noted that, unlike instructors of business or law, instructors of science have rarely used the case study method (1). Fortunately, that is changing as instructors of various scientific disciplines are adapting this method to their courses. As a teacher of biochemistry, I became interested in the case study method because I believed that the problem-solving skills developed by the students during these exercises would be of more benefit to them in their future endeavors than mindless memorization of endless pathways. So a few years ago I began to include a single case study exercise in an out-of-class assignment. The assignment was given after the students had completed the intermediary metabolism portion of the course. The class was divided into small groups, and each group was given a different case to solve. Since most of my students are interested in pursuing medicine, I gave them clinical cases. In preparing these assignments I found a rich source of material on clinical cases. I often consulted Stanbury's classic textbook *The Metabolic Basis of Inherited Disease* (2) or several textbooks devoted to the case-oriented approach (3-4) or books containing compilations of cases (5-7).

The students' responses to the case study exercise were very positive. They enjoyed interacting in small groups and sharing their knowledge of biochemistry (in fact, some were surprised at how much they had learned!) They enjoyed seeing that biochemistry was "relevant" in a "real-life" situation. And several of the students wrote on evaluation forms at the end of the semester "Why don't we do case studies all semester long?"

Why not, indeed? I found the answer to that question as I searched for materials on biochemistry cases that were not related to medicine and found that there were no such resources. Thus I began to write my own cases. As my library of case studies grew, I expanded the number of case studies from one to five during the semester, and dropped the more traditional homework assignments. In addition, on four occasions during the semester we have an in-class case study discussion in place of the regular lecture. I happened to mention these activities casually to Carl Beers, the John Wiley & Sons representative for the area including Providence College, and the result is this casebook.

The case studies presented here are exercises that I've written using current journals in the field as my primary resources. I have tried to use current studies whenever possible to maintain student interest. The case topics chosen are those found in most biochemistry textbooks on the market. Each case includes a focus concept, prerequisites, a background summary and a series of questions. The prerequisites include biochemistry material that the students are likely to be studying in class, but occasionally include concepts such as genetics and immunology on a level likely to be covered in a first-year general biology course.

My students have responded positively to these assignments, and I have found that they learn the material as well as or better than if they had completed a more traditional assignment. I am grateful to John Wiley & Sons for giving me the chance to share these cases with you, and I would welcome any comments or suggestions for improvement.

Kathleen Cornely
Providence College
November, 1998

1. Herreid, C. F. (1994) "Case Studies in Science--A Novel Method of Science Education" *Journal of College Science Teaching* **23**(4), pp. 221-229.

2. Scriver, C. R., Beaudet, A. L., Sly, W. S., and Valle, D., eds. (consulting eds. Stanbury, Wyngaarden and Fredrickson) *The Metabolic Basis of Inherited Disease*, 1989, McGraw-Hill.

3. Montgomery, R., Conway, T. W., and Spector, A. *Biochemistry: A Case-Oriented Approach* (1990), The C. V. Mosby Company, St. Louis.

4. Devlin, T. M. (ed.) *Textbook of Biochemistry with Clinical Correlations* (1992) Wiley-Liss, NY.

5. Halperin, M. L., and Rolleston, F. S. *Clinical Detective Stories: A Problem-Based Approach to Clinical Cases in Energy and Acid Base Metabolism* (1993) Portland Press, London.

6. Lorenza, R. F. *Learning Biochemistry: 100 Case Oriented Problems* (1995) Wiley-Liss, NY.

7. Higgins, S. J., Turner, A. J., and Wood, E. J. (1994) *Biochemistry for the Medical Sciences: An Integrated Case Approach*, Wiley & Sons, NY.

Acknowledgments

I would like to thank the following individuals for their help while I was preparing this casebook:

I owe a debt of gratitude to the wonderful folks at John Wiley & Sons for their assistance on this project: Carl Beers, for recommending me for the project; Jennifer Yee and the Wiley staff, for providing several of the drawings that appear in this casebook, and Cliff Mills, who is the best editor that anyone could ever hope to have.

I would also like to thank Angela Medici of Studio Medici, Pawtucket, RI, for assistance with the layout of the casebook.

The library staff at Providence College, especially Frances Mancini, were especially helpful in obtaining the many references I needed to complete the casebook.

I would like to thank W. Scott Champney, East Tennessee State University; Gary L. Powell, Clemson University; and Larry L. Jackson, Montana State University for reviewing an early draft of the manuscript. I am also indebted to Barbara Brennessel, Wheaton College; Koni Stone, California State University-Stanislaus; and Kristin Wobbe, Worcester Polytechnic Institute, who carefully critiqued the final draft and made suggestions that resulted in a much improved manuscript.

I would also like to thank the faculty and staff of the Chemistry Department at Providence College who have always provided support for me professionally and personally. I am also grateful to the College administration who supported my attendance at several case study conferences.

My students at Providence College have provided helpful suggestions and insights and served as "guinea pigs" while I was preparing these cases. Special thanks go to Amy Bouchard '00, Anthony Denis '00, Joseph Matrullo '00 and Paul Simoes '00 for helpful comments on Case 1, Lauren Ayr '00 for suggestions on Case 10, and to Beth Ploszay '97 for giving me the idea for Case 9.

And finally to my husband David Moss for providing editorial advice, helpful suggestions, and many dinners!

About the Author

Kathleen Cornely is an Associate Professor in the Department of Chemistry at Providence College in Providence, Rhode Island, where she teaches biochemistry lecture and laboratory courses, in addition to courses in general and organic chemistry. She holds a B.S. degree in chemistry from Bowling Green (Ohio) State University, an M.S. in biochemistry from Indiana University, and a Ph.D. in nutritional biochemistry from Cornell University. Her primary research interests are in the area of protein purification and chemical modification, with applications including the action of the peptide hormone glucagon, the role of hepatic lipase in lipid metabolism, and the isolation of a membrane-bound protein believed to help maintain the asymmetry of lipid bilayer membranes. She has also published several manuscripts on educational issues in biochemistry and general chemistry.

Her interest in case studies grew out of her choice to include them in a one-semester biochemistry course, both as a means of making biochemical problems more interesting to her students and as a method of illustrating the interconnectedness of biochemical systems. She has participated in case study workshops sponsored by the Pew Midwest Consortium and by the University of Buffalo (through a grant from the National Science Foundation).

Contents

Case 1
Acute Aspirin Overdose:
Relationship to the Blood Buffering System

Focus concept

The response of the carbonic acid/bicarbonate buffering system to an overdose of aspirin is examined.

Prerequisites

- Principles of acids and bases, including pK_a and the Henderson-Hasselbalch equation.
- The carbonic acid/bicarbonate blood buffering system.

Background

You are an emergency room physician and you have just admitted a patient, Susan M., a 22-year-old female, who was brought to the emergency room around 9 pm by her friend Anne S. Anne tells you that she had stopped by Susan's apartment and found that Susan was disoriented and had trouble speaking. Anne brought Susan to the emergency room when Susan began to suffer from nausea and vomiting. She is also hyperventilating. Anne reveals that Susan had been depressed lately, and shows you an empty aspirin bottle she had found in Susan's apartment. The aspirin bottle, when full, would have contained 250 tablets. Susan admits that she took the tablets around 7 pm that evening. You draw blood from Susan and the laboratory performs the analyses shown in Table 1.1.

Table 1.1: Arterial blood gas concentration in patient Susan M.

	Susan M., two hours after aspirin ingestion	Susan M., ten hours after aspirin ingestion	Normal values
P_{CO_2}	26 mm Hg	19 mm Hg	35-45 mm Hg
HCO_3^-	18 mM	21 mM	22-26 mM
P_{O_2}	113 mm Hg	143 mm Hg	75-100 mm Hg
pH	7.44	7.55	7.35-7.45
Blood salicylate concentration, mg/dL	57	117	

In the emergency room, Susan is given a stomach lavage with saline and two doses of activated charcoal to adsorb the aspirin. Eight hours later, she was still experiencing nausea and vomiting, and her respiratory rate increased, and further treatment was required. You carry out a gastric lavage with a sodium bicarbonate solution, pH = 8.5. Susan's blood pH begins to drop around 24 hours after the aspirin ingestion and finally returns to normal at 60 hours after the ingestion.

Questions

1. Aspirin, or acetylsalicylic acid (structure shown in Figure 1.1), is converted in the stomach in the presence of aqueous acid and stomach esterases (which act as catalysts) to salicylic acid and acetic acid. Write the balanced chemical reaction for this transformation.

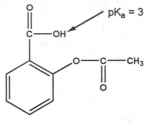

Acetylsalicylic acid (aspirin)

Figure 1.1: Structure of aspirin.

2. Since Anne has brought Susan into the emergency room only two hours after the overdose, you suspect that Susan's stomach might contain undissolved aspirin that is continuing to be absorbed. The fact that Susan is still experiencing nausea and vomiting 10 hours after the ingestion confirms your suspicion and you decide to use a gastric lavage at pH 8.5 to effectively remove the undissolved aspirin. This treatment solubilizes the aspirin so that it can easily be removed from the stomach.

   a. Calculate the percentage of protonated and unprotonated forms of salicylic acid at the pH of the stomach, which is usually around 2.0.

   b. Calculate the percentage of protonated and unprotonated forms of salicylic acid at the pH of the gastric lavage. Why does the gastric lavage result in increased solubility of the drug? (Note: Assume that the pK_a values for the carboxylate group in salicylic acid and acetylsalicylic acid are the same.)

3. Describe how H_2CO_3/HCO_3^- serves as a buffer, using relevant equations.

4. It has been shown that salicylates act directly on the nervous system to stimulate respiration. Thus, our patient is hyperventilating due to her salicylate overdose. Explain how the salicylate-induced hyperventilation leads to the symptoms seen in our patient. Use the appropriate equations and cite and explain the relevant laboratory data.

5. Use the Henderson-Hasselbalch equation to determine the ratio of HCO_3^- to H_2CO_3 in the patient's blood 10 hours after aspirin ingestion. The pK_a of the first dissociable proton of H_2CO_3 is 6.4. How does this compare to the ratio of HCO_3^- to H_2CO_3 in normal blood? Can the H_2CO_3/HCO_3^- system serve as an effective buffer in this patient? Explain.

6. Sixty hours after aspirin ingestion, the patient's blood pH has returned to normal (pH = 7.4). Describe how the carbonic/bicarbonate buffering system responded to bring the patient's blood pH back to normal.

7. Are there other substances in the blood that can serve as buffers?

Reference

Krause, D. S., Wolf, B. A., and Shaw, L. M. (1992) *Therapeutic Drug Monitoring* **14**, pp. 441-451.

Case 2
Histidine-Proline-rich Glycoprotein as a Plasma pH Sensor

Focus concept

A histidine-proline-rich glycoprotein may serve as a plasma sensor and regulate local pH in extracellular fluid during ischemia or metabolic acidosis.

Prerequisites

- Acidic/basic properties of amino acids.
- Amino acid structure and protein structure.

Background

This study focuses on the characteristics of the abundant plasma protein referred to as histidine-proline-rich glycoprotein (HPRG). HPRG is so named because it has a very high content of histidine (13 mol%). The human HPRG contains a pentapeptide GHHPH sequence repeated in tandem twelve times. The authors of this study hypothesized that its high histidine content might allow HPRG to play a role in regulating local pH in the blood. The local pH in blood may drop a half a pH unit during lactic acidosis or even a full pH unit in hypoxia or ischemia. In the case presented here, the binding of HPRG to glycosaminoglycans was investigated. Glycosaminoglycans are anionic polysaccharides that are the major component of the ground substance that forms the matrix of the extracellular spaces of the connective tissue in blood vessel walls. In this study, the binding of HPRG to the glycosaminoglycan heparin was measured. Based on their results, the investigators propose a model which describes how binding of HPRG to glycosaminoglycans may allow HPRG to regulate local blood pH.

Questions

1. Binding studies were carried out in which heparin was immobilized on the surface of cuvettes. The pH dependence of HPRG binding to heparin was investigated and the results are shown in Figure 2.1.
   a. How is the binding of HPRG to heparin dependent on pH? Give structural reasons for the binding dependence. The structure of heparin is shown in Figure 2.2.
   b. The same binding studies were carried out in which HPRG was reacted with diethylpyrocarbonate (DEPC), a compound that specifically reacts with histidine residues. The reaction is shown in Figure 2.3. Explain the results.

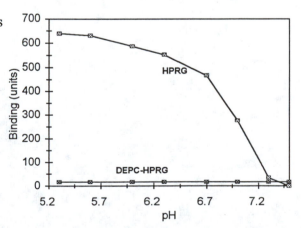

Figure 2.1: The pH-dependence of HPRG binding to unmodified heparin, and heparin modified with DEPC. (Based on Borza and Morgan, 1998.)

3

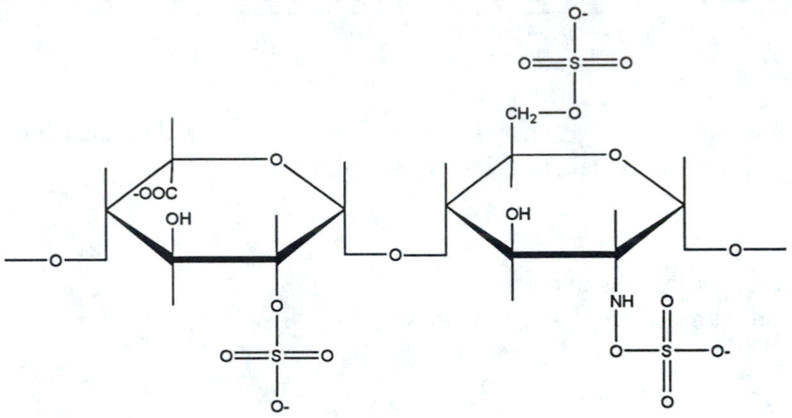

Figure 2.2: Repeating disaccharide unit of heparin.

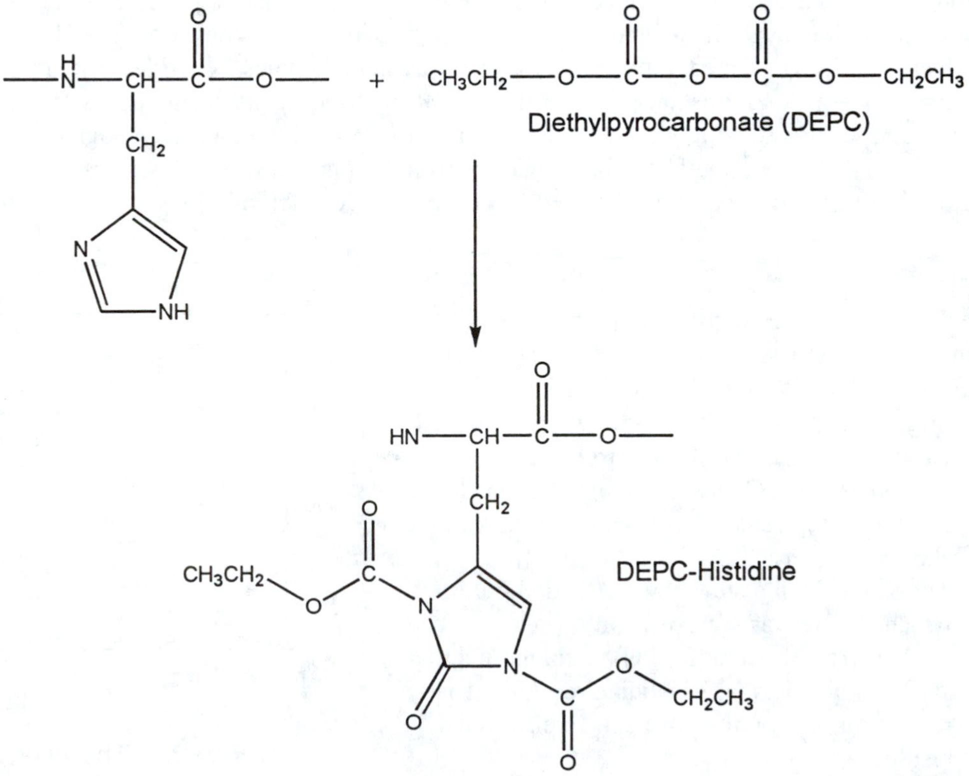

Figure 2.3: Reaction of histidine side chains with diethylpyrocarbonate.

2. The effect of transition metals on the binding of HPRG was investigated next. The ability of increasing concentrations of Cu^{2+} and Zn^{2+} to promote HPRG binding to heparin at pH = 7.3 was measured. The results are shown in Figure 2.4. In addition, the binding of HPRG to heparin in the presence of these ions was compared at various pH's. Figure 2.5 shows the comparison of binding at pH = 6.0 and at pH = 7.4 in the presence of 5.2 nM zinc. What is your interpretation of these results?

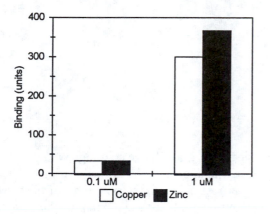

Figure 2.4: Binding of HPRG to heparin in the presence of copper and zinc ions. (Based on Borza and Morgan, 1998.)

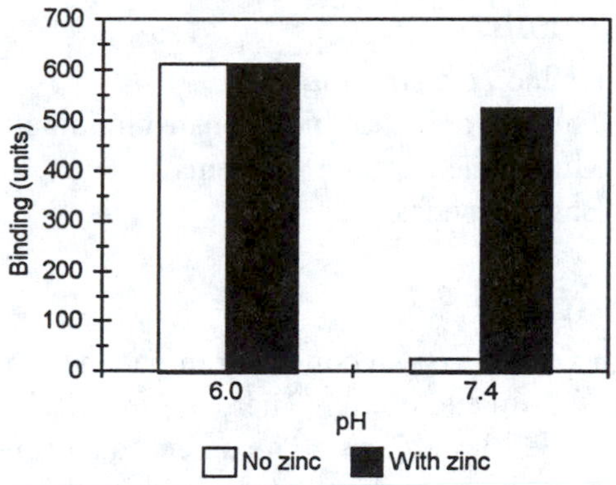

Figure 2.5: Binding of HPRG to heparin in the presence and absence of 5.2 nM zinc ions at two different pH values. (Based on Borza and Morgan, 1998.)

3. Local cellular pH can decrease from one-half to one pH unit depending on a variety of circumstances including ischemia, hypoxia, and inflammation due to lactic acidosis. In addition, metabolic acidosis is often one of the symptoms in complications following surgery. The investigators have proposed that HPRG acts to relieve the acidosis in these circumstances. Propose a model that explains the mechanism of pH regulation by HPRG.

4. Other plasma proteins have been studied for their ability to bind to glycosaminoglycans. One such protein is kininogen, which is a lysine-rich protein. Like HPRG, kininogen is able to bind to glycosaminoglycans, but this binding is far less sensitive to small fluctuations in physiological pH.
   a. Why does kininogen bind to glycosaminoglycans easily?
   b. Why is the binding of kininogen less sensitive to physiological pH changes?

References

Borza, D-B., and Morgan, W. T. (1998) *J. Biol. Chem.*, **273**, pp. 5493-5499.

Lundblad, R. (1995) *Techniques in Protein Modification*, CRC Press, Boca Raton, FL, p. 111.

Case 3
Carbonic Anhydrase II Deficiency

Focus concept

The role of the carbonic anhydrase enzyme in normal bone tissue formation is examined.

Prerequisites

- Amino acid structure.
- The carbonic acid/bicarbonate blood buffering system.
- Membrane transport proteins.
- Basic genetics.

Background

In this case, we will consider our patients: three sisters, aged 21, 24, and 29 years of age who are short of stature and obese. (There is a fourth sister in the family who appears to be normal, as she is taller than the other three sisters. The parents also appear to be normal.) As children, the symptoms of the three sisters were similar–delayed mental and physical development, muscle weakness, and renal tubular acidosis. They frequently suffered bone fractures as children. X-rays showed cerebral calcification and other skeletal abnormalities. After reviewing the sisters' medical histories, you draw samples of blood and send it to the laboratory for analysis. The laboratory reports to you that your patients all have a carbonic anhydrase II deficiency.

There are seven isozymes of carbonic anhydrase (CA), three of which occur in humans and are designated CA I, II and III. They are all monomeric zinc metalloenzymes and have molecular weights of 29 kilodaltons. X-ray crystallographic data shows that the enzyme is roughly spherical with the active site located in a conical cleft. One side of this cleft is lined with hydrophobic amino acid residues while the other side is lined with hydrophilic residues. The zinc ion is located at the bottom of the cleft and is coordinately covalently bound to the imidazole rings of three histidine residues.

The carbonic anhydrase II isozyme is found in bone, kidney, and brain, which is why the defects occur in these tissues when the enzyme is deficient or non-functional. The carbonic anhydrase II enzyme is highly active, with one of the highest turnover rates of any known enzyme, and is critical in maintaining proper acid-base balance.

Questions

1. Carbonic anhydrase catalyzes the reaction between water and carbon dioxide to yield carbonic acid. The carbonic acid then undergoes dissociation. Write the two equations that describe these processes. What products form when carbonic acid is dissociated?

2. Each of the three sisters with the symptoms described above showed a carbonic anhydrase II deficiency. In contrast, the fourth sister and both parents showed half-normal levels of the enzyme. Construct a chart which describes how the carbonic anhydrase deficiency syndrome is inherited. Note that the defective carbonic anhydrase gene is inherited as an autosomal recessive gene.

3. A genetic analysis of one of the sister's genes indicates that a (His → Tyr) mutation at amino acid 107 is responsible for the carbonic anhydrase deficiency. Using what you know about amino acid structure, propose a hypothesis that might explain why such a mutation would result in an inactive enzyme.

4. Osteoclasts in bone tissue are particularly rich in carbonic anhydrase II, and a proper functioning enzyme is critical to the development of healthy tissue. In order for proper bone development to occur, the osteoclast must acidify the bone-resorbing compartment. Also involved in this acidification are several transporters: a Na^+/H^+ exchanger, a Cl^-/HCO_3^- exchanger and the Na^+K^+ATPase, which exchanges Na^+ and K^+ ions. (An *exchanger* is a protein or protein complex located in the cell membrane which transports one ion in one direction and the second ion in the other direction simultaneously.)

A partial diagram of the osteoclast is shown in Figure 3.1. Fill in the blanks in the diagram indicating the roles of carbonic anhydrase II and the exchangers in the acidification of the bone-resorbing compartment. Include the reactants and products of the appropriate intracellular reaction(s) and note in which direction each ion is transported in the osteoclast.

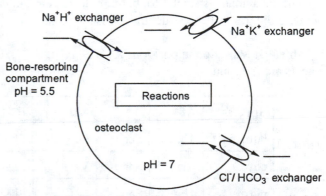

Figure 3.1: The role of the osteoclast intracellular carbonic anhydrase II in establishing the acidity of the bone-resorbing compartment.

References

Sly, W. S., and Hu, P. Y. (1995) *Ann. Rev. Biochem.*, **64**, pp. 375-401.

Whyte, M. P. (1993) *Clin. Orthop. Relat. Res.*, **294**, pp. 52-63.

Case 4
The Structure of Insulin

Focus concepts
The three dimensional structure of insulin is examined and sequences of various animal insulins are compared.

Prerequisites
- Amino acid structure.
- Protein architecture.
- Basic immunology.

Background
Diabetics lack the protein insulin, which is produced by the pancreatic β-cells of the islets of Langerhans. Insulin stimulates uptake of glucose from the blood into the tissues. Diabetes is treated by replacing the missing insulin. Human insulin is produced industrially by recombinant bacteria, but before this method was available, animal insulin was used instead.

Insulin consists of two polypeptide chains, an A chain and a B chain, joined together by disulfide bonds. The smaller of the two chains is referred to as the A chain and is 21 amino acids long in humans. The second chain is referred to as the B chain and is 30 amino acids long in humans. Insulin from various animals is similar to, but not identical to human insulin, as illustrated in Table 4.1. A schematic diagram of the structure of insulin is shown in Figure 4.1.

Table 4.1: Variation in positions A8, A9, A10, B1, B2, B27 and B30 of insulin. (All other amino acids are the same.)

Species	A8	A9	A10	B1	B2	B27	B30
human	Thr	Ser	Ile	Phe	Val	Thr	Thr
cow	Ala	Ser	Val	Phe	Val	Thr	Ala
pig	Thr	Ser	Ile	Phe	Val	Thr	Ala
horse	Thr	Gly	Ile	Phe	Val	Thr	Ala
rabbit	Thr	Ser	Ile	Phe	Val	Thr	Ser
dog	Thr	Ser	Ile	Phe	Val	Thr	Ala
chicken	His	Asn	Thr	Ala	Ala	Ser	Ala
duck	Glu	Asn	Pro	Ala	Ala	Ser	Thr

Questions

1. What animals would serve as the best sources of insulin to be used for treating diabetics? Explain your answer.

2. Would the pI values of the animal insulins be the same as, greater than, or less than human insulin?

3. Some people developed allergies to the animal insulin because their immune systems recognized the proteins as foreign. Explain why the immune system would be able to distinguish animal insulin from human insulin.

4. An SDS-PAGE gel is run of proinsulin and insulin. Samples were treated with β-mercaptoethanol prior to electrophoresis. Draw a picture of the predicted results.

5. A denaturation/renaturation (similar to the one carried out by Anfinsen with ribonuclease) experiment was carried out using insulin. However, in contrast to Anfinsen's results, only less than 10% of the activity of insulin was recovered when urea and β-mercaptoethanol were removed by dialysis. (This is the level of activity you would expect if the disulfide bridges paired randomly.) In contrast, if the experiment is repeated with proinsulin, full activity is restored upon renaturation. Explain these observations.

6. Diabetics are treated with insulin, not proinsulin. Do you think that this is a good idea?

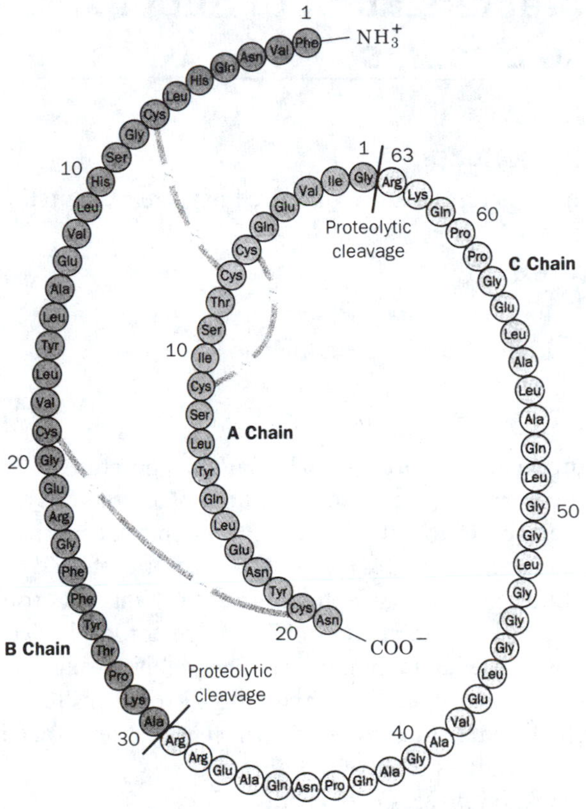

Figure 4.1: Structure of proinsulin. (From Voet and Voet, 1995.)

References

Devlin, T. M., ed. (1997) *Textbook of Biochemistry with Clinical Correlations*, Wiley-Liss, NY, p. 41.

Steiner, D. F., and Rubenstein, A. H. (1997) *Science* **277**, pp. 531-532.

Voet, D., and Voet, J. (1995) *Biochemistry*, Second Edition, John Wiley and Sons, New York p.191-194.

Characterization of Subtilisin from the Antarctic Psychrophile *Bacillus* TA41

Focus concept

The structural features involved in protein adaptation to cold temperatures are explored.

Prerequisites

- Protein architecture.
- Principles of protein folding.

Background

Although most organisms live at a temperature of 37°C, some organisms have the ability to survive at extreme temperatures of heat and cold. Bacteria referred to as thermophiles can survive in hot springs at temperatures up to 120°C. Other organisms such as bacteria found in Antarctic seawater are capable of surviving at temperatures below 0°C and are referred to as psychrophiles. Growth at such extreme temperatures requires adaptation of the organism's proteins so that enzymatic reactions will still be able to take place. For the thermophiles, a greater level of stability is required so that the proteins will not denature at the high temperature. This stability is achieved through ion pairs, hydrophobic interactions, and ion binding. In addition, thermophilic proteins tend to be compact because they have fewer hydrophilic amino acid residues which have a preference for interacting with the solvent. Adaptation at cold temperatures is slightly different because additional stability is not needed. Instead, because reaction rates typically decrease with decreasing temperature, psychrophilic enzymes must compensate by being able to accommodate their substrates easily so that the reaction can take place in a timely manner. This requires proteins with additional conformational flexibility and decreased stability.

The properties of thermophilic proteins have been extensively studied, but not as much is known about their psychrophilic counterparts. In this case, the investigators purified and characterized the protease subtilisin S41 from the Antarctic

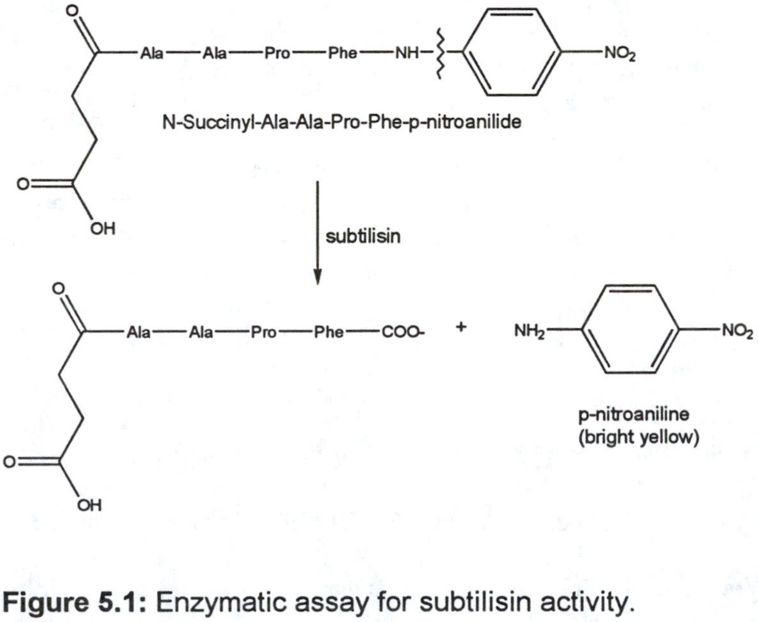

Figure 5.1: Enzymatic assay for subtilisin activity.

psychrophile *Bacillus* TA41. Subtilisin was chosen because the properties of nearly 50 subtilisin-like proteins are known. Subtilisin S41 is secreted by the *Bacillus* into its growth medium, making isolation of this enzyme a fairly straightforward process. Once purified, the subtilisin S41 was sequenced and characterized. The activity of the subtilisin protease enzyme was measured by adding the substrate N-succinyl-Ala-Ala-Pro-Phe-*p*-nitroanilide (AAPF) and measuring the absorbance of the yellow *p*-nitroaniline product at 412 nm. The reaction is shown in Figure 5.1. The properties of the psychrophilic subtilisin were then compared with subtilisin from mesophiles, organisms that grow at more moderate temperatures.

Understanding the factors that allow a protein to adapt to a cold environment may provide clues to the protein folding process.

Questions

1. Following purification of the subtilisin S41 enzyme, the investigators carried out analytical procedures to determine the protein's molecular weight and pI.
   a. What technique would they have used to determine the protein's molecular weight? What technique would they have used to determine the protein's pI?
   b. The pI of the psychrophilic subtilisin S41 enzyme was determined to be 5.3. This value is much lower than the pI of mesophilic subtilisins, which have pI values ranging from 9-11. Which amino acid residues does the subtilisin S41 have in greater amounts than the mesophilic subtilisins?

2. The investigators then carried out an experiment in which they incubated subtilisin S41 and subtilisin Carlsberg (a mesophilic subtilisin) at 50°C and measured enzymatic activity as a function of time. The results are shown in Table 5.1. Interpret these results and explain their significance.

Table 5.1: Thermal stability of subtilisin S41 and subtilisin Carlsberg. Enzymes were incubated at 50°C and enzyme activity was measured at periodic intervals. The half-times of inactivation, t_i, (time required to cause a 50% reduction in enzyme activity) are reported, as well as the optimum temperature of activity.

Enzyme	t_i, minutes	Optimal temperature of activity, °C
Subtilisin S41	3	40
Subtilisin Carlsberg	24	60

3. The properties of psychrophilic, mesophilic, and thermophilic subtilisin enzymes were compared and the results are shown in Table 5.2. Using these data and data in Question 2, describe the general characteristics of psychrophilic subtilisin, and compare these characteristics with the mesophilic and thermophilic subtilisins.

Table 5.2: Main structural features of subtilisins are compared. (Based on Davail, *et al.*, 1994.)

	Psychrophilic	Mesophilic			Thermophilic
	S41	*BPN*	*Carlsberg*	*Savinase*	*Thermitase*
Amino Acids	309	275	274	269	279
Asp Content	21	10	9	5	13
Ionic Interactions	2	5	3	7	10
Aromatic Inter-actions	0	5	3	7	10
K_d for Ca^{2+} (M)*	10^{-6} M	10^{-10} M	10^{-10} M	not known	$< 10^{-10}$ M

* The K_d value is the Ca^{2+} concentration required to achieve 50% calcium binding to the subtilisin.

Reference
Davail, S., Feller, G., Narinx, E., and Gerday, C. (1994) *J. Biol.Chem.*, **269**, pp. 17448-17453.

Case 6
Ehlers Danlos Syndrome Type VII:
Collagen Synthesis Disorder

Focus concept

Connective tissue disorders result if collagen cannot be synthesized properly. In this case, experimental data is examined to pinpoint the exact nature of the defective synthesis in a patient with Ehlers Danlos Syndrome Type VII.

Prerequisites

- Structural properties of collagen.
- The collagen biosynthetic pathway, including the role of Vitamin C.
- General properties of enzymes.

Background

Your patient in this case is Linda T., a 10-year-old girl of short stature with hyperelastic skin which bruises easily, and hypermobile joints which are easily dislocated. Linda's mother mentions to you that she had taken a biochemistry course some years before and that she thought that Linda might have a Vitamin C deficiency. She changed Linda's diet to include fruits and vegetables high in Vitamin C, but this failed to correct the problem.

You decide to refer Linda to a surgeon who will correct the laxity of Linda's joints. The surgeon agrees to take skin samples at the time that the surgery is performed so that you can perform an analysis of her fibroblasts. Because skin, ligaments, and tendons have collagen as their major structural protein, you suspect that Linda is unable to synthesize collagen properly and you wish to test this hypothesis.

That evening, you have dinner with one of your old college friends who is now a veterinarian. You describe Linda's symptoms to her and she remarks that Linda's symptoms sound remarkably similar to a condition called dermatosparaxis (literally "torn skin") that she has observed in sheep and cattle. She agrees with you that the disorder probably is due to improper collagen formation, and generously offers a sample of collagen from a dermatosparactic sheep that you can use in your experiments. You are able to obtain fibroblasts from the samples of Linda's skin and the skin from the dermatosparactic sheep, as well as fibroblasts capable of secreting normal collagen.

Collagen is synthesized in fibroblast cells in the skin. There are four main types of collagen, but you suspect that the defect is in Type I collagen synthesis, because Type I collagen is predominantly distributed in tendons, bone, skin, and ligaments. Type I collagen consists of three chains: two identical $\alpha 1(I)$ chains and an $\alpha 2$ chain. The biosynthetic pathway is shown in Figure 6.1.

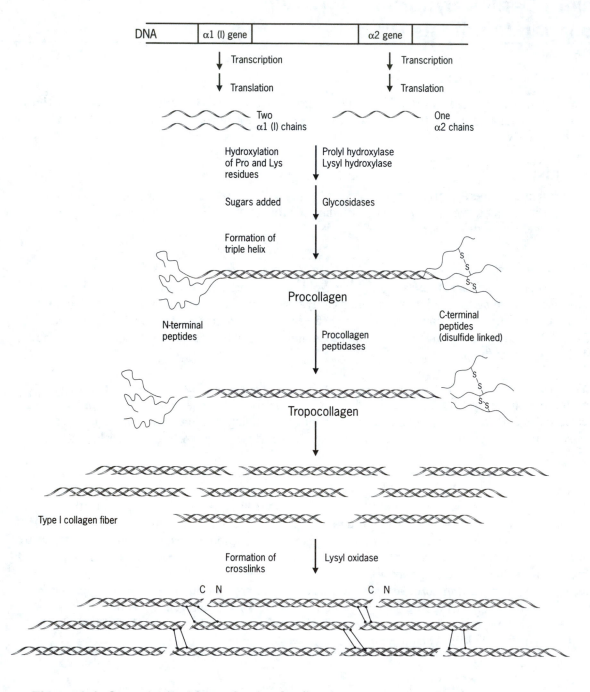

Figure 6.1: Steps in the biosynthesis of collagen.

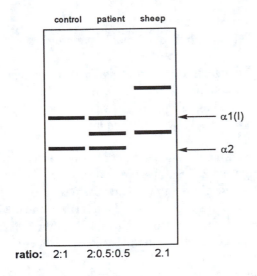

control   patient   sheep

← α1(I)

← α2

ratio: 2:1   2:0.5:0.5   2.1

Figure 6.2: SDS-PAGE analysis of collagen isolated from control, patient, and sheep fibroblasts. The ratios indicated at the bottom of the gel are the ratios of intensities of the bands (from top to bottom) as determined by densitometric scanning. (Based on Steinman, *et al.*, 1980.)

Table 6.2: Experiments with collagen taken from an Ehlers Danlos Syndrome Type VII patient and a dermatosparactic sheep.

	Collagen extracted from Linda's fibroblasts	Collagen extracted from dermatosparactic sheep
Assays of enzyme activity		
Procollagen peptidases		
N-protease	normal	low
C-protease	normal	normal
Prolyl hydroxylase	normal	normal
Lysyl hydroxylase	normal	normal
Relative amino acid compositions		
4-Hydroxyproline (4-Hyp)	normal	normal
3-Hydroxyproline (3-Hyp)	normal	normal
Hydroxylysine (Hyl)	normal	normal
Incubation with exogenous enzyme		
Procollagen N–protease	No change	N-terminal propeptides cleaved on both α1(I) and α2 chains
Procollagen C-protease	No change	No change

Questions

1. Why did increasing the amount of Vitamin C in Linda's diet fail to improve Linda's symptoms?

2. In the first step of your analysis, you extract and solublize collagen from fibroblasts from the patient, the sheep, and the control. You find that it's much easier to obtain collagen from the patient's fibroblasts than from either the sheep or control fibroblasts. Using what you know about the collagen biosynthetic pathway make a list of the possible defects in Linda's Type I collagen biosynthesis.

3. In the next experiment, you carry out SDS-PAGE analysis of the collagen extracted from normal, patient, and sheep fibroblasts. The results are shown in Figure 6.2. You observe that there is an extra band in Linda's collagen that has a slightly higher molecular weight than the molecular weight of the $\alpha2$ chain. In the sheep there are two bands with molecular weights greater than the $\alpha1(I)$ and the $\alpha2$ chains. Do these results help you narrow down your list of possibilities from Question 2? Is the defect in collagen synthesis in Linda's fibroblasts different from the defect in the dermatosparactic sheep?

4. Explain the significance of the ratios of collagen chains in the control and patient collagen as shown in the SDS-PAGE gel in Figure 6.2.

5. Next, you carry out a series of experiments outlined in Table 6.2. In the first set of experiments, the fibroblasts are assayed for levels of key enzymes necessary in proper collagen synthesis. In the second set of experiments, an amino acid analysis of control and patient collagen is carried out. In the third set of experiments, exogenous enzymes are added to the cultured fibroblast medium to see if these enzymes can correct the defect. The results are shown in the table. Does this help your narrow down the list of possibilities? Is the defect in Linda's collagen synthesis different from the defect in dermatosparaxis? Explain how the defect results in fragile collagen in both the patient and the dermatosparactic sheep.

References

Stanbury, J. B., Wyngaarden, J. B., and Frederickson, D. S., eds. (1978) *The Metabolic Basis of Inherited Disease*, McGraw-Hill Book Company, New York, pp. 1368-1379.

Steinmann, B., Tuderman, L., Peltonen, L., Martin, G. R., McKusick, V. A., and Prockop, D. J. (1980) *J. Biol. Chem.*, **255**, pp. 8887-8893.

A Storage Protein From Seeds of *Brassica nigra* is a Serine Protease Inhibitor

Focus concept

Purification of a novel seed storage protein allows sequence analysis and determination of the protein's secondary and tertiary structure.

Prerequisites

- Protein purification techniques, particularly gel filtration and dialysis.
- Protein sequencing using Edman degradation and overlap peptides.
- Structure and mechanism of serine proteases.
- Reversible inhibition of enzymes.

Background

Seedlings use seed storage proteins as an important nitrogen source during germination. The seed storage proteins are made as large precursors, then hydrolyzed to smaller products for the seedling's use during growth. In this case, the investigators discovered a new seed storage protein, which they named BN, in the oilseed *Brassica nigra*. These seeds are important nutritionally as a source of oil as well as protein. The storage protein described here was first purified and then characterized for its important biochemical properties. The storage protein turned out to be an inhibitor of serine protease enzymes. The authors hypothesized that the purpose of serine protease inhibitors like BN is to protect the plant from proteolytic enzymes of insects and microorganisms that would damage the plant.

Questions

1. In order to isolate the protein, seeds were ground and extracted with water. The proteins in the extract were precipitated with hydrochloric acid, isolated by centrifugation, then lyophilized (freeze-dried). The powder was dissolved in a small amount of ammonium acetate buffer at pH = 5 and the sample was loaded onto a Sephadex gel filtration column. The elution profile showed four peaks. Most of the BN protein eluted in the first peak.
   a. On what basis is separation achieved on the Sephadex gel filtration column?
   b. What statement can you make about the BN protein relative to other proteins found in the *B. nigra* seeds?

2. Following gel filtration chromatography, the BN protein was further purified by dialysis using tubing with a 6000-8000 molecular weight cut-off. Analysis using SDS-PAGE showed a single band with a molecular weight of 15,500 daltons. Why does dialysis yield a more highly purified protein?

3. Next, the investigators attempted to sequence the protein using an Edman degradation procedure. However, this was initially unsuccessful because the amino terminus was blocked. Based on comparisons with other proteins in the same family as BN whose sequences are known, the investigators hypothesized that the amino terminal amino acid was N-acetyl serine.
   a. Draw the structure of N-acetyl serine.
   b. If N-acetyl serine was indeed the amino terminal amino acid, why would sequencing using the Edman method be unsuccessful?

4. The amino acid sequence of BN was finally determined in the following manner: The BN protein was first treated with β-mercaptoethanol to reduce any disulfide bridges. This treatment revealed that the BN protein consisted of two chains, a light chain and a heavy chain. Separation of the two chains by disulfide bridge reduction destroyed the inhibitory capabilities of the BN protein. Next, the light chain and heavy chain were separated and then three separate samples of the purified chains were treated with three different proteases. The fragments obtained were individually sequenced using the Edman method. The protein sequence is shown in Table 7.1. The light chain is 39 amino acids long and the heavy chain is 91 amino acids long.
   a. Why was it necessary to carry out a minimum of two different proteolytic cleavages of the protein using different proteases?
   b. One of the enzymes used by the investigators was trypsin. Write the sequences of the fragments that would result from trypsin digestion.
   c. Choose a second protease to cleave both the light and heavy chains into smaller fragments. What protease did you choose, and why? Write the sequences of the fragments that would result from the digestion of the protease you chose.

Table 7.1: Amino acid sequences of the BN protease inhibitor from *Brassica nigra* seeds. Note that the first five amino acid residues of the light chain are missing due to a blocked amino terminal amino acid. (Based on Genov, *et al.*, 1997.)

Light chain

	1	2	3	4	5	6	7	8	9	10	11	12	13	14	15
1						Arg	Ile	Pro	Lys	Cys	Arg	Lys	Glu	Phe	Gln
16	Gln	Ala	Gln	His	Leu	Arg	Ala	Cys	Gln	Gln	Trp	Leu	His	Lys	Gln
31	Ala	Asn	Gln	Ser	Gly	Gly	Gly	Pro	Ser						

Heavy chain

	1	2	3	4	5	6	7	8	9	10	11	12	13	14	15
1	Pro	Gln	Gly	Pro	Gln	Gln	Arg	Pro	Pro	Leu	Leu	Gln	Gln	Cys	Cys
16	Asn	Glu	Lys	His	Gln	Glu	Glu	Pro	Leu	Cys	Val	Cys	Pro	Thr	Leu
31	Lys	Gly	Ala	Ser	Lys	Ala	Val	Arg	Gln	Gln	Ile	Arg	Gln	Gln	Gly
46	Gln	Gln	Gln	Gly	Gln	Gln	Gly	Gln	Gln	Leu	Gln	Arg	Glu	Ile	Ser
61	Arg	Ile	Tyr	Gln	Thr	Ala	Thr	His	Leu	Pro	Arg	Val	Cys	Asn	Ile
76	Pro	Arg	Val	Ser	Ile	Cys	Pro	Phe	Gln	Lys	Thr	Met	Pro	Gly	Pro
91	Ser														

5. Circular dichroism (CD) is a technique used for determining the secondary structure (α-helix and β-pleated sheet) of a protein. The authors took CD spectra of the protein at varying temperatures. The results are shown in Figure 7.1. What do these data tell you about the structure of the protein at various temperatures?

6. The BN protein is a competitive inhibitor of the serine proteases trypsin, subtilisin and chymotrypsin. The percentage inhibition was measured for each enzyme in the presence of BN protein. The results are shown in the Table 7.2. Using what you know about enzyme inhibition, describe how BN inhibits the proteolytic activity of these serine proteases. Be sure to explain how a single inhibitor can inhibit three different enzymes.

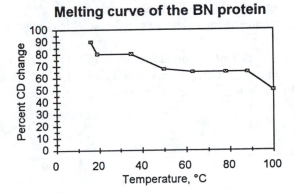

Figure 7.1: Single wavelength melting curve of the protease inhibitor from *Brassica nigra*. (Based on Genov, *et al.*, 1997)

Table 7.2: The BN protein is a serine protease inhibitor. (Based on Genov, *et al.*, 1997)

Enzyme	% inhibition at [BN] = 2.0 x 10⁻⁶ M
trypsin	100%
subtilisin	100%
chymotrypsin	32%

7. The authors next carried out fluorescence studies in order to obtain additional information concerning the three-dimensional structure of the BN protein. Only tryptophan and tyrosine are *fluorophores*, which means they are capable of undergoing fluorescence. They found that a single tryptophan residue emits light at 330 nm upon excitation at 295 nm.

There are several ions and small molecules that have the ability to *quench* fluorescence. This means that upon excitation at 295 nm, the tryptophan will transfer its energy to the quenching agent rather than releasing the energy in emitted light. Cesium (Cs^+) ions, iodide (I^- ions) and acrylamide (see structure in Figure 7.2) are capable of quenching the tryptophan residue's fluorescence–that is, if these quenching agents can make suitable contact with the tryptophan.

The investigators found that there was no quenching with cesium ions. Iodide and acrylamide demonstrated low quenching efficiencies, although acrylamide's efficiency was higher than iodide's. What do these observations tell you about the microenvironment of the tryptophan residue?

$$CH_2 = CH - \overset{\overset{\displaystyle O}{\|}}{C} - NH_2$$

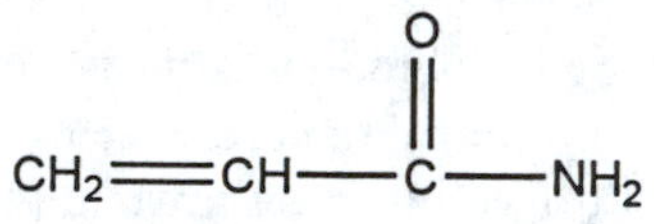

Figure 7.2: Structure of acrylamide.

Reference

Genov, N., Goshev, I., Nikolova, D., Georgieva, D. N., Filippi, B., and Svendsen, I. (1997) *Biochim. Biophys. Acta*, **1341**, pp. 157-164.

Case 8
Hemoglobin, the Oxygen Carrier

Focus concept

A mutation in the gene for hemoglobin results in an altered protein responsible for the disease sickle cell anemia. An understanding of the biochemistry of the disease may suggest possible treatments.

Prerequisite

Hemoglobin structure and function concepts.

Background

Normal adult hemoglobin is called Hemoglobin A (Hb A). Ninety-eight percent of adult hemoglobin is Hb A and 2% is Hb A_2. There are other forms of hemoglobin. For example, the developing fetus has a different kind of hemoglobin than most normal adults. Fetal hemoglobin (or Hemoglobin F) consists of two α chains and two γ chains, whereas adult hemoglobin (Hemoglobin A) consists of two α chains and two β chains. Fetal hemoglobin is synthesized beginning at the third month of gestation and continues up through birth. After the neonate is born, hemoglobin F synthesis declines (because synthesis of the γ chain declines) and hemoglobin A is synthesized (because synthesis of β chains begins). By the time the baby is six months old, 98% of its hemoglobin is Hemoglobin A.

There is also a mutant form of hemoglobin called Hemoglobin S which is found in persons with the disease sickle cell anemia. The disease sickle cell anemia is one of the major health problems facing the African-American community. The World Health Organization estimates that 250,000 babies world wide are born with sickle cell anemia. Currently there is no cure. A person afflicted with sickle cell anemia has inherited a defective gene from each parent. (Parents who are carriers of the sickle cell gene are heterozygous AS, whereas the person afflicted with sickle cell anemia is SS; non-carriers are designated AA.) The defective gene is the one coding for the β-chain. The amino acid at position 6 on each β chain has been mutated from a glutamate to a valine. Red blood cells containing Hb S form a sickle shape because the Hb S molecules clump together. Hb S molecules are more likely to clump together when in the deoxygenated T form than in the oxygenated R form. The sickle shaped red blood cells become trapped in capillaries and organs, depriving the victim of adequate oxygen supply and causing chronic pain and organ damage.

In this case we will consider our patient, a 10-year-old black male child named Michael B., who was admitted to the hospital because he was experiencing severe chest pain. He had been hospitalized on several previous occasions for vaso-occlusive episodes that caused him to experience severe pain that could not be managed with non-prescription drugs such as ibuprofen. He was slightly jaundiced, short of breath and easily tired, and feverish. A chest x-ray was taken and was abnormal. An arterial blood sample showed a p_{O2} value of 6 kPa (normal is 10-13 kPa).

21

Questions

1. You suspect that Michael has sickle cell anemia and you have ordered an isoelectric focusing analysis of the child's lysed red blood cells. (Lysing the red blood cells releases the hemoglobin.) Draw a diagram of the predicted results. Why will this test allow you to diagnose this child's disease?

2. Why do you think that Hb S molecules would be likely to clump together whereas Hb A molecules do not?

3. In the emergency room, oxygen (100%) was administered to the patient. (Inspired air normally is about 20% oxygen.) Why was this an effective treatment?

4. You recall reading in the medical literature about a dramatic new drug treatment for sickle cell anemia, and you'd like to try it on this patient. The drug is hydroxyurea, and is thought to function by stimulating the afflicted person's synthesis of fetal hemoglobin. In a clinical study, patients who took hydroxyurea showed a 50% reduction in frequency of hospital admissions for severe pain, and there was also a decrease in the frequency of fever and abnormal chest x-rays. Why would this drug alleviate the symptoms of sickle cell anemia?

5. A year ago, at a conference, one of your colleagues told you that she had "cured" a patient of sickle cell anemia by performing a bone marrow transplant. Why would this procedure "cure" sickle cell anemia?

6. The patient's parents tell you that they are planning on having another child and that they are confident that subsequent children will not have sickle cell anemia, since they already have a child with the disease. What will you tell them?

Reference

Glew, R. H., and Ninomiya, Y. (1997) *Clinical Studies in Medical Biochemistry*, Oxford University Press, pp. 78-90.

Case 9
Allosteric Interactions in Crocodile Hemoglobin

Focus concept

The effect of allosteric modulators on oxygen affinity for crocodile hemoglobin is unique when compared with other species.

Prerequisite

Hemoglobin structure and function concepts.

Background

While most human beings are able to hold their breath for only a minute or two, other species are able to stay under water for much longer periods of time. In this case study we will examine the physiological adaptations that allow some organisms to deliver oxygen to tissues while submerged under water.

Deep sea-diving mammals, such as whales and seals, are able to stay under water for long periods of time. These mammals are able to stay submerged because their muscles contain many-fold higher concentrations of myoglobin (Mb) than humans.

Crocodiles are also able to stay submerged under water for periods of time exceeding one hour. This adaptation allows the crocodile to kill small mammals by drowning them. However, the crocodile doesn't have large amounts of myoglobin in its muscle as the deep sea-diving mammals do, so their physiological adaptation must be different. In 1995, Nagai and colleagues described in the British journal *Nature* a possible mechanism that allowed the crocodile hemoglobin to deliver a large fraction of bound oxygen to the tissues. They suggested that bicarbonate, HCO_3^-, binds to hemoglobin to promote the dissociation of oxygen in a manner similar to 2,3-bisphosphoglycerate (BPG) in humans.

This case is important because information gathered from experiments like those described here will allow scientists to design effective blood replacements.

Questions

1. In humans, oxygen is effectively delivered to the tissues because of the presence of several allosteric modulators. Name three of these modulators and explain how their presence allows oxygen to be delivered to the tissues.

2. Explain why having higher concentrations of Mb would allow whales and seals to stay submerged under water for a long period of time.

3. Let us consider the hypothesis that bicarbonate serves as an allosteric modulator of hemoglobin binding in crocodiles. What is the source of HCO_3^- in the crocodile tissues?

4. Draw oxygen-binding curves for crocodile hemoglobin in the presence and absence of bicarbonate. Which conditions give rise to a greater p_{50} value for crocodile hemoglobin? What does this tell you about the oxygen binding affinity for hemoglobin under those conditions?

5. Komiyama *et al.* investigated the bicarbonate binding site on the crocodile hemoglobin by constructing human-crocodile chimeric hemoglobins in which amino acids in the human hemoglobin were replaced with amino acids found in the crocodile hemoglobin at the same location. (The investigators wanted to see if they could make a synthetic human hemoglobin that resembled the crocodile hemoglobin in terms of its ability to bind bicarbonate anions.) They found the bicarbonate binding site to be located at the $\alpha_1\beta_2$-subunit interface, where the two subunits slide with respect to one another during R $\rightleftarrows$ T transitions. Based on their results the authors modeled a *stereochemically plausible* binding site that included the phenolate anion of Tyr 41β, the ε-amino group of Lys 38β, and the phenolate anion of Tyr 42α.

   What kinds of interactions do you think the aforementioned amino acid side chains will have with the bicarbonate anion? (It might be helpful to draw the Lewis electron dot structure of bicarbonate).

6. In order to create an engineered human hemoglobin molecule that had the same bicarbonate binding properties as crocodile hemoglobin, twelve amino acid residues had to be changed. Not all of these residues directly interact with bicarbonate–perhaps only three of them do, as described in question 5. What might be the role of the other nine amino acid residues?

7. Other animals have similarly adapted to using small molecules as allosteric effectors to encourage hemoglobin to release its oxygen. Whereas humans use 2,3-BPG and crocodiles use HCO_3^-, birds use *myo*-inositol pentaphosphate (IP_5) and fish use ATP and GTP. The structures of ATP and IP_5 are shown in Figure 9.1. What structural characteristics do all of these molecules have in common and how would they bind to hemoglobin?

Reference

Komiyama, N. H., Miyazaki, G., Tame, J., and Nagai, K. (1995) *Nature* **373**, pp. 244-246.

ATP

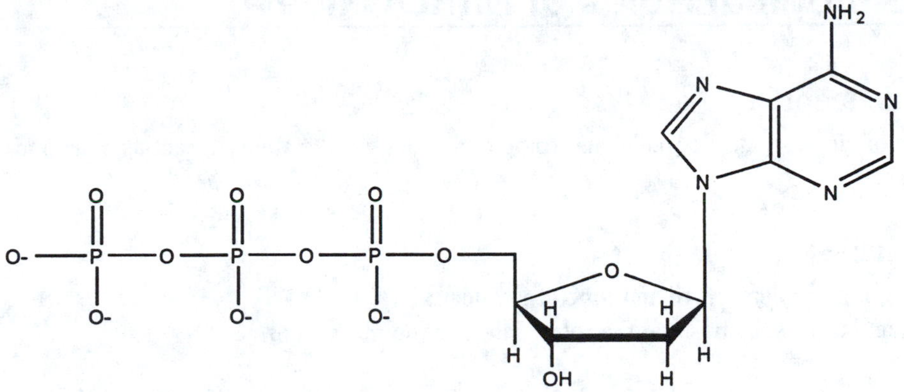

GTP

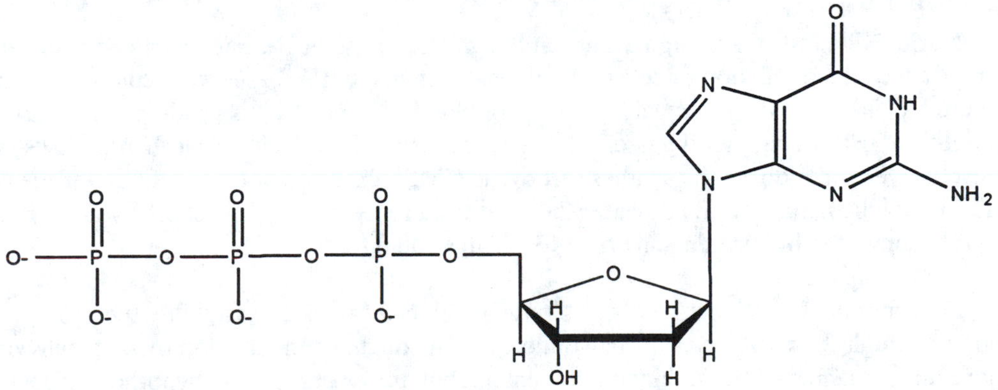

Myo-Inositol Pentaphosphate

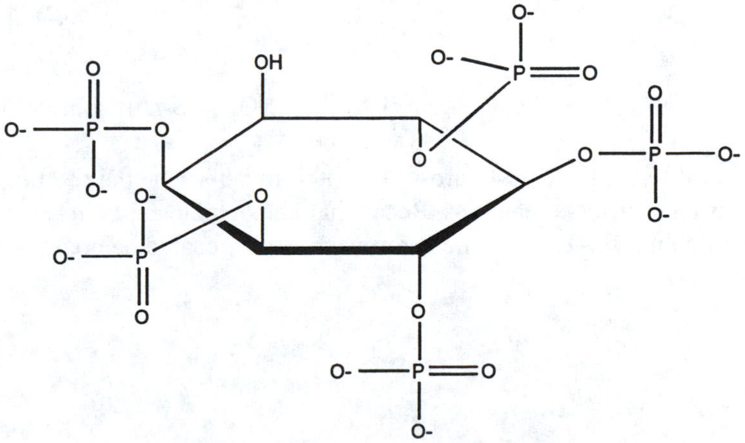

Figure 9.1: Allosteric effectors of hemoglobin in various species.

25

The Biological Roles of Nitric Oxide

Focus concept

Nitric oxide, a small, lipophilic and toxic molecule, has the ability to act as a second messenger in blood vessels.

Prerequisites

- Hemoglobin structure and function concepts.
- Regulatory schemes and the role of the second messenger.

Background

Nitric oxide (NO), although a small and highly toxic molecule, has recently been found to be important in the regulation of several biological systems. In this case, we focus on the circulatory system where NO is one of the principal regulators of blood pressure. NO is produced from arginine in endothelial cells lining blood vessels, then released from the cells where it then diffuses into the adjacent muscle layer and induces the synthesis of cyclic GMP which promotes relaxation of the blood vessel. (The lipophilic nature of NO obviates the need for a receptor.) NO is short-lived (5-10 seconds) and is rapidly converted by oxygen and water to nitrates and nitrites.

In promoting the relaxation of the blood vessel, NO can clearly facilitate the passage of oxygenated blood through the vessel. Scientists wondered if NO might be involved in oxygen delivery to the cells in a more direct manner–through binding to hemoglobin, for example. This hypothesis was confirmed recently when investigators showed that NO can react with hemoglobin to form S-nitrosohemoglobin. The NO reacts with the sulfhydryl group on Cys 93 on the β chain of hemoglobin to form the S-nitroso-hemoglobin. The reaction is readily reversible, with SNO-Hb releasing the NO under certain circumstances.

$$\text{hemoglobin (Hb)} \quad + \quad \text{NO} \quad \rightleftarrows \quad \text{S-nitrosohemoglobin (SNO-Hb)}$$

In this case, the investigators attempted to determine the reaction conditions favorable for the forward and reverse reactions. Recall that hemoglobin exists in two different conformations: Oxygenated hemoglobin (Oxy-Hb) has the R form and deoxygenated hemoglobin (deoxy-Hb) has the T form.

26

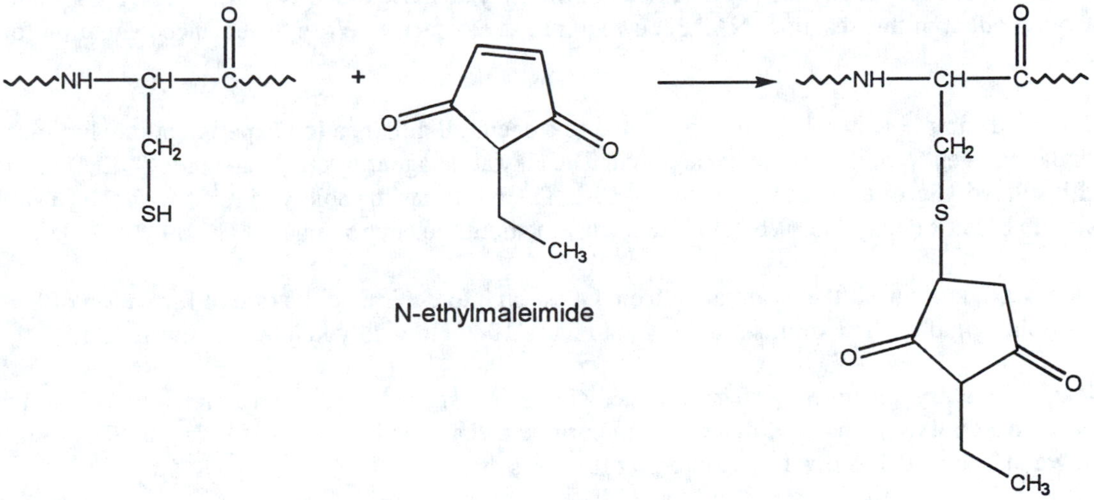

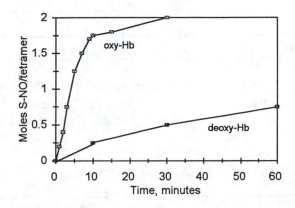

Figure 10.1: Modification of cysteine residues in proteins with the alkylating agent **N**-ethylmaleimide.

Figure 10.2: Rate of reaction of oxy-Hb and deoxyHb with NO to form S-nitroso-hemoglobin. (Based on Jia, *et al.*, 1996.)

Questions

1. NO is produced from arginine by the action of the enzyme nitric oxide synthase (NOS). Citrulline is also a product in this reaction. NADPH is required as a cofactor. Write the balanced equation for the reaction.

2. A purified sample of hemoglobin is subjected to a chemical modification experiment in which an alkylating agent is added to the hemoglobin. The alkylating agent N-ethylmaleimide (NEM) reacts with sulfhydryl groups as shown in Figure 10.1. The NEM has the ability to react with the Cysβ93 in oxy-Hb but not deoxy-Hb. What does this tell you about the environment of Cysβ93?

3. The investigators used the information from Question 2 to determine the rate of formation of S-nitrosohemoglobin. The results are shown in Figure 10.2. How do you interpret these data?

4. Next, circulatory dynamics were investigated. The scientists collected arterial and venous blood from anesthetized rats. The hemoglobin was then purified and assayed for SNO-Hb content. The results are shown in Table 10.1. How do you interpret these results?

Table 10.1: Endogenous concentrations of S-nitrosohemoglobin in arterial and venous blood. (Based on Jia, *et al.*, 1996.)

Blood	SNO-Hb (nM)	
	No NOS inhibitor	With NOS inhibitor
Arterial	311	82
Venous	32	not measured

5. Next the investigators injected purified hemoglobin and purified SNO-Hb into the circulation of an experimental animal, then measured the animal's blood pressure. The results are shown in Table 10.2. How do you interpret these results? (Assume that the injection of the purified proteins did not affect blood volume.)

Table 10.2: Reactivity of Hb and SNO-Hb with endogenous NO. (Based on Jia, *et al.*, 1996.)

	Hb	SNO-Hb
Change in arterial pressure	↑ 20 mm Hg	No change

6. Oxygen delivery to capillaries is therefore controlled by two factors: The oxygen content of the hemoglobin molecule itself, and the ability of the blood vessel to relax so that the blood can flow through the capillary easily. Through the action of NO, both of these processes are linked. The investigators proposed that the S-nitrosohemoglobin somehow senses the oxygen demand in a particular tissue and acts accordingly.

    a.   Describe how oxygen is delivered effectively to cells in hypoxic (oxygen-deprived) tissue. For example, what would be the relative concentrations of SNO-Hb and Hb? What is the effect on the blood vessels?

    b.   What would the relative concentrations of SNO-Hb and Hb be in blood circulated through tissue in which oxygen was plentiful? What is the effect on blood vessels in this situation?

References

Blakeslee, S. (July 22, 1997) "What Controls Blood Flow? Blood" *The New York Times*, p. C1.

Hsia, C. C. W. (1998) *New Engl. Jour. Med.* **338**, pp. 239-247.

Jia, L., Bonaventura, C., Bonaventura, J., and Stamler, J. S. (1996) *Nature* **380**, pp. 221-226.

Snyder, S. H., and Bredt, D. S. (1992) *Scientific American*, **266(5)**, pp. 68-77.

Stamler, J. S., Jia, L., Eu, J. P., McMahon, T. J., Demchenko, I. T., Bonaventura, J., Gernert, K., and Piantadosi, C. (1997) *Science* **276**, pp. 2034-2037.

Case 11
Nonenzymatic Deamidation of Asparagine and Glutamine Residues in Proteins

Focus concept

Factors influencing nonenzymatic hydrolytic deamidation of Asn and Gln residues in proteins are examined and possible mechanisms for the reactions are proposed.

Prerequisites

- Protein analytical methods, particularly isoelectric focusing.
- Enzyme mechanisms, especially proteases such as papain and chymotrypsin.

Background

Asparagine (Asn) and glutamine (Gln) residues in proteins will sometimes be nonenzymatically hydrolytically deamidated to aspartate (Asp) and glutamate (Glu) residues, respectively. An exhaustive study of proteins known to undergo deamidation has revealed that the rate of deamidation is influenced by the amino acids preceding and following the amide amino acids in primary sequence. Other amino acids that might be far apart in primary sequence to the deamidated amino acids, but close in terms of tertiary structure, might also influence the deamidation process.

The extent to which deamidation occurs might be underestimated. Proteins purified using traditional biochemical techniques frequently provide an environment in which deamidation can occur readily. Since Asp and Glu residues resulting from deamidation are indistinguishable from Asp and Glu residues originally present in the protein, the proper sequence of the protein is not always known. Thus, the only way to know whether a protein undergoes deamidation is to compare the sequences of bases in the gene to the amino acid sequence of the purified protein. Unfortunately, these data are not available for every protein that undergoes deamidation.

The deamidation of amide side chains results in a change in the tertiary structure of the protein. In fact, it has been suggested that deamidation actually functions as a molecular timer for protein turnover, since it has been observed that deamidation of some proteins increases their susceptibility to degradation by proteases. Proteins that turn over rapidly might do so as a result of a rapid deamidation process which would signal cellular proteases to destroy the deamidated protein. In the same vein, deamidation might be involved in the aging process. This hypothesis is best tested in proteins such as the lens crystallin protein that turn over slowly or not at all. It has been shown in chickens that 15% of the lens crystallin protein is deamidated at age four months, whereas 50% of the protein is deamidated at one year and 70% at ten years.

Questions

1. Asparagine and glutamine residues in proteins are deamidated in an aqueous solution to yield aspartate and glutamate, respectively, and ammonium. Write the balanced chemical equations for these processes.

2. Proteins undergoing deamidation of one or more of their Asn or Gln groups have been detected by isoelectric focusing. Why would isoelectric focusing be effective in separating deamidated from non-deamidated protein? Compare the relative pI values of the deamidated and non-deamidated proteins.

3. The primary amino acid sequences were examined for proteins known to undergo hydrolytic deamidation of their Asn residues. The results are shown in Figure 11.1. What statements can you make concerning the kinds of amino acids you would expect to find before and after the labile Asn?

4. The mechanism of deamidation of the amide side chain involves the participation of an acid catalyst, shown in Figure 11.2 as HA.
   a. Propose a mechanism for the deamidation process. The first step is provided in Figure 11.2. The transition intermediate should be shown.
   b. Look at the structure of the transition intermediate. What amino acid side chains might participate in stabilizing this structure? Be specific about the kinds of noncovalent interactions involved.

5. Since the deamidation of Asn and Gln residues is known to be non-enzymatic, the HA acid catalyst cannot be provided by an enzyme. Instead, the *catalytic* groups are believed to be provided by neighboring amino acids in the protein undergoing deamidation. Refer to your answer to Question 3 and examine the mechanism you have just written. Describe how amino acid side chains that either precede or follow the labile Asn could serve as catalytic groups in the deamidation process.

6. Amino terminal Gln residues are particularly susceptible to deamidation, and undergo deamidation much more rapidly than internal Gln residues. In the deamidation process, a five-membered pyrrolidone ring structure is formed.
   a. Write a mechanism for this deamidation process.
   b. Amino terminal Asn residues do not undergo deamidation. Why do you think this is the case?

7. The tertiary structure of a protein must also be considered when comparing deamidation rates. Another source of catalytic groups are amino acid side chains such as the ones discussed above that are located in the same vicinity as the labile Asn or Gln. Another consideration is the location of the Asn and Gln with respect to the tertiary structure of the protein. It has been observed that Asn and Gln residues in the interior of a protein are deamidated at a much slower rate than Asn and Gln residues on the surface of the protein. Why is this the case?

8. Explain how deamidation could affect the tertiary structure of a protein. Be specific about what kinds of non-covalent interactions might be involved.

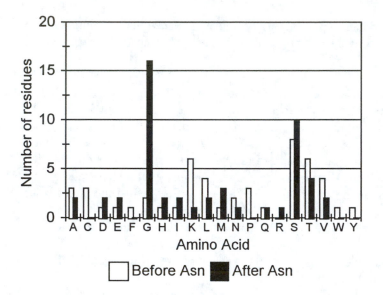

Figure 11.1: Frequency with which each of the twenty amino acids occurs before and after the labile Asn residues in a set of proteins known to undergo nonenzymatic deamidation. (Based on Wright, 1991.)

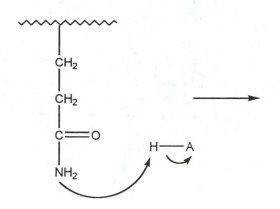

Figure 11.2: The first step of a proposed mechanism for the hydrolytic deamidation of Gln. An acid catalyst, labeled HA, participates in the reaction. (Based on Wright, 1997.)

References

Robinson, A. B., and Rudd. C. (1974) *Curr. Top. Cell Regul.*, **8**, 247-295.

Robinson, A. B., Scotchler, J. W., and McKerrow, J. H. (1973) *J. Am. Chem. Soc.* **95**, 8156-8159.

Robinson, A. B., McKerrow, J. H., and Cary, P. (1970) *Proc. Natl. Acad. Sci. USA* **66**, 753-757.

Wright, H. T. (1991) *Critical Reviews in Biochemistry and Molecular Biology* **26**, 1-52.

Case 12
Production of Methanol in Ripening Fruit

Focus concept

The link between the production of methanol in ripening fruit and the activity of pectin methyl-esterase, the enzyme responsible for methanol production, is examined in wild type and transgenic tomato fruit.

Prerequisite

Enzyme kinetics and inhibition.

Background

Methanol is produced by fruit during the ripening process. The major source of this methanol is likely to be from the enzyme pectin methylesterase (PME) which acts upon pectin methyl esters to de-esterify them and produce methanol and other pectic substances. One might assume that the production of methanol is correlated with PME activity, but this has not been firmly established. In this case, the investigators at the Plant Science Department at Rutgers University developed a transgenic tomato which was deficient in PME activity. By comparing the behavior of the transgenic tomato with the wild type tomato, the production of methanol in the ripening tomato fruit may be better understood.

Questions

1. The investigators measured PME enzyme activity in immature green (IMG), mature green (MG), breaker (Br), turning (Tu) and red ripe (RR) tomatoes. They also determined the amount of PME protein present by using a Western blot assay, which uses a specific antibody to detect the PME protein. The results are shown in Figure 12.1. What is your interpretation of these results?

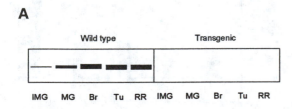

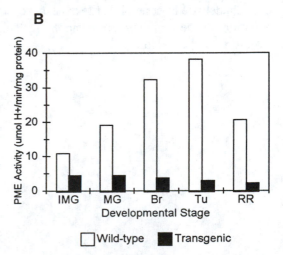

Figure 12.1: PME protein levels (A) and activity (B) in wild-type and transgenic tomatoes. (Modified from Frenkel, *et al.*, 1998.)

2. The investigators next wanted to study the correlation between PME activity and methanol levels. They measured methanol content of tomato tissue at each developmental stage of ripening as described in Question 1. The results are shown in Figure 12.2.

   a. Compare the results in Figure 12.2 with those presented in Figure 12.1. Is methanol production correlated with PME activity? Be specific.

   b. What is responsible, at least in part, for the production of methanol in the ripening fruit?

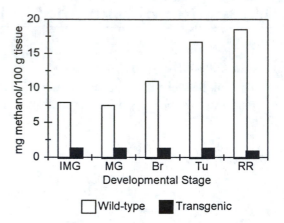

Figure 12.2: Methanol content in ripening wild-type and transgenic tomato fruit. (Based on Frenkel, *et al.*, 1998.)

3. Next, the investigators measured the ethanol content in the ripening fruit for the wild type and transgenic tomatoes. The results are shown in Figure 12.3. Compare the results in Figure 12.2 with the results presented in Figure 12.3. What is your interpretation of these results?

4. Methanol is produced from the action of PME on pectin methyl esters, as described above. The only route for ethanol production in the plant is the alcohol dehydrogenase-catalyzed reduction of acetaldehyde to ethanol. Write a balanced equation for the production of ethanol, including structures and cofactors.

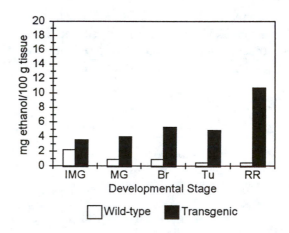

Figure 12.3: Ethanol content in ripening wild-type and transgenic tomato fruit. (Based on Frenkel, *et al.*, 1998.)

34

5. In order to understand the mechanism for ethanol production, the investigators measured the activity of the alcohol dehydrogenase (ADH) enzyme in both the wild type and transgenic tomatoes at each stage of the fruit-ripening process. The results are shown in Figure 12.4. What factor or factors might influence ADH activity in the ripening tomato?

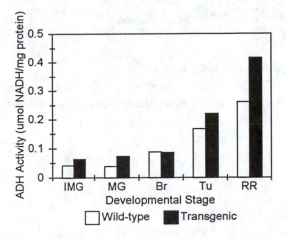

Figure 12.4: Alcohol dehydrogenase activity in ripening wild-type and transgenic tomato fruit. (Based on Frenkel, *et al.*, 1998.)

Reference

Frenkel, C., Peters, J. S., Tieman, D. M., Tiznado, M. E., and Handa, A. K. (1998) *J. Biol. Chem.*, **273**, pp. 4293-4295.

Case 13
A "Flippase" Enzyme Maintains Membrane Asymmetry

Focus concept

The properties of a phospholipid translocase enzyme are determined via a series of experiments with sonicated phospholipid vesicles and red blood cells.

Prerequisites

- Phospholipid structure and nomenclature.
- Singer-Nicholson fluid mosaic model of a membrane.

Background

It has been well established that the cellular phospholipid membrane bilayer is asymmetric. Choline-containing lipids such as phosphatidylcholine (PC) and sphingomyelin (SM) are more likely to be found in the outer leaflet of the bilayer whereas aminophospholipids like phosphatidylserine (PS) and phosphatidylethanolamine (PE) are more likely to be found in the inner leaflet of the bilayer. Several investigators have used a variety of methods in order to determine what mechanism is used by the cell to maintain this asymmetry. Several studies have indicated that a transmembrane protein that has been nicknamed a "flippase" is likely to be partially responsible for maintaining phospholipid asymmetry in the bilayer. It has been hypothesized that the flippase enzyme would act by "flipping" a phospholipid from the outer leaflet of the membrane to the inner leaflet. Specific transport of some types of phospholipids but not others would create the asymmetric distribution that has been observed.

In this case, the investigators took solutions of a single kind of phospholipid and sonicated the solutions to form phospholipid vesicles. The vesicles were then added to red blood cells. The phospholipids in the vesicle migrated from the vesicle to the red blood cells. The investigators then used a microscope to observe the shape of the red blood cells. Red blood cells normally have a biconcave disk shape. If an excess amount of phospholipid is added to the outer membrane leaflet, the red blood cells are said to become crenated or echinocytic or "spiky". But if the added phospholipid is acted upon by the flippase enzyme, then the phospholipid would be flipped from the outer leaflet to the inner leaflet of the membrane. This results in an excess of phospholipids in the inner leaflet of the membrane. As a result, the red blood cell surface would be covered with "craters" and is said to be stomatocytic. By observing which phospholipids could be flipped and which could not, the investigators were able to ascertain the properties of the flippase enzyme. In this case, we will examine some of these experiments and then use our observations to describe the flippase enzyme in more detail. The experimental design is shown in Figure 13.1.

In order to quantitate their results, the investigators used a numbering system that is diagramed in Figure 13.2. Biconcave red blood cell disks were given a score of zero. If the red blood cell became echinocytic, it was given a score from +1 to +5, indicating an increasing number of "spikes." Similarly, stomatocytic cells were given scores from -1 to -4, indicating an increasing number of "craters". In each experiment, a field of 100 cells was observed, each cell was assigned numbers, and the numbers were averaged to yield a value called the morphological index (MI). Echinocytic cells have an excess amount

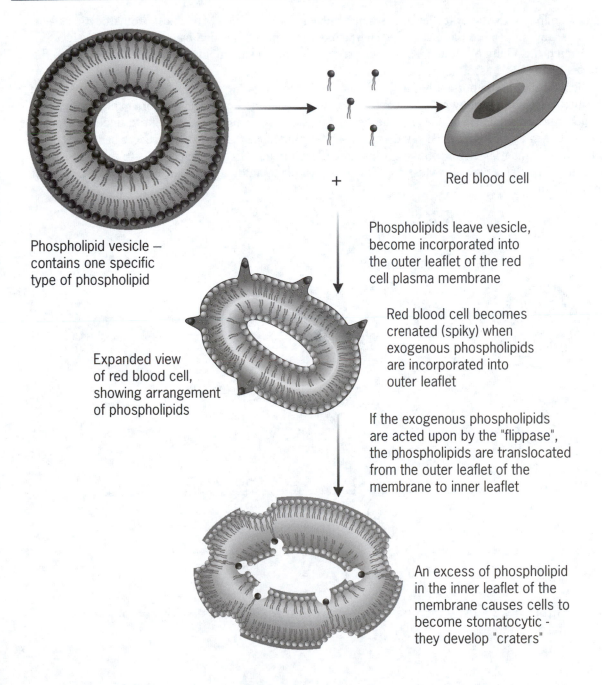

Phospholipid vesicle –
contains one specific
type of phospholipid

Red blood cell

Phospholipids leave vesicle,
become incorporated into
the outer leaflet of the red
cell plasma membrane

Red blood cell becomes
crenated (spiky) when
exogenous phospholipids
are incorporated into
outer leaflet

Expanded view
of red blood cell,
showing arrangement
of phospholipids

If the exogenous phospholipids
are acted upon by the "flippase",
the phospholipids are translocated
from the outer leaflet of the
membrane to inner leaflet

An excess of phospholipid
in the inner leaflet of the
membrane causes cells to
become stomatocytic -
they develop "craters"

Figure 13.1: Experimental protocol for shape change experiments.

of phospholipid in the outer leaflet of the membrane and have positive MI values. Stomatocytic cells have an excess amount of phospholipid in the inner leaflet in the membrane and have negative MI values. In this manner, the ability of the flippase enzyme to translocate phospholipids from the outer leaflet to the inner leaflet could be ascertained.

These studies are important because it is known that the asymmetric distribution of lipids just described is found in healthy cells. Under some circumstances, however, the membrane phospholipids may become "scrambled" and the asymmetry will be lost. This occurs in cells about to undergo a process called *apoptosis*, or programmed cell death. An understanding of the apoptotic process has implications for the development of therapeutic drugs that might encourage cells like cancer cells to undergo apoptotic death. This case also illustrates how it is possible to study the properties of a protein without purifying it first.

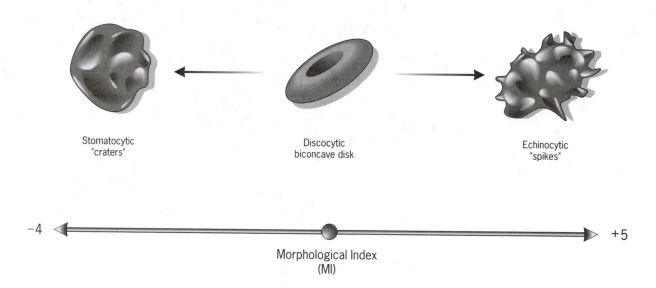

Figure 13.2: The morphological index (MI) scale for assessing red blood cell shape change.

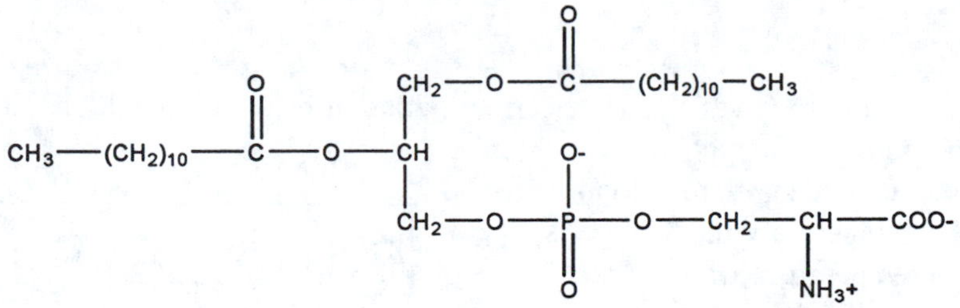

Dilauroylphosphatidylserine (DLPS)

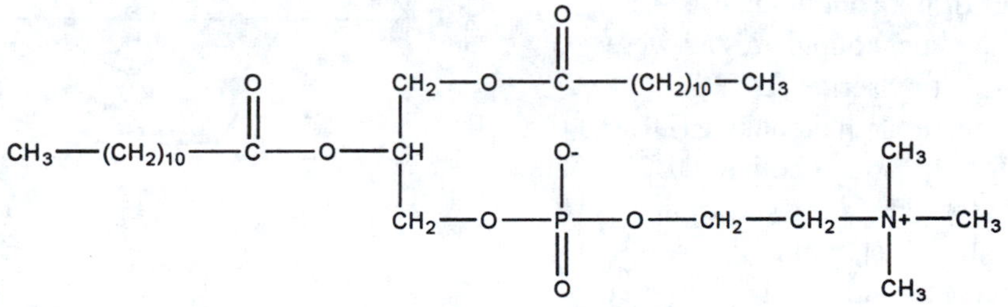

Dilauroylphosphatidylcholine (DLPC)

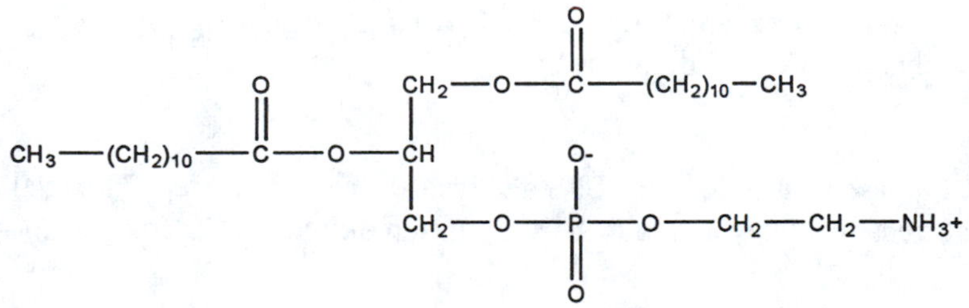

Dilauroylphosphatidylethanolamine (DLPE)

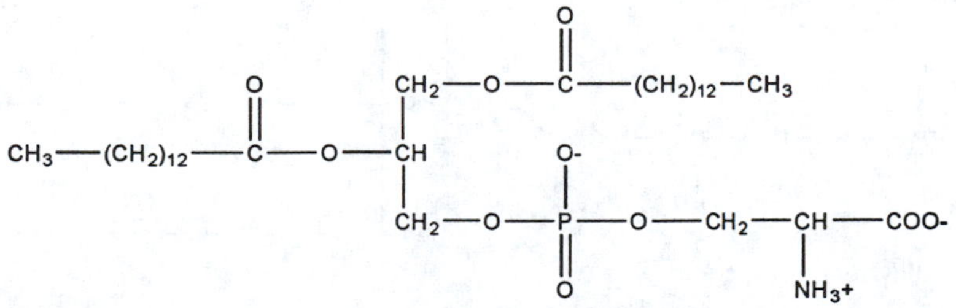

Dimyristoylphosphatidylserine (DMPS)

Figure 13.3: Structures of phospholipids used in shape change experiments.

Questions

1. Experiments with phospholipid vesicles were carried out as shown in Figure 13.1. First, phospholipid vesicles containing one specific type of phospholipid were incubated with red blood cells at 4°C (at this temperature the flippase is inactive). Then the temperature was warmed up to 37 °C to activate the flippase. The ability of the flippase to translocate three different types of phospholipids was measured. The phospholipids tested were dilauroylphosphatidylserine (DLPS), dilauroylphosphatidylethanolamine (DLPE) and dilauroylphosphatidylcholine (DLPC). The structures of these lipids are shown in Figure 13.3. The results are shown in Figure 13.4.

   a. What type(s) of phospholipids is(are) preferentially translocated by the flippase? Explain.

   b. Given the distribution of lipids in the membrane, what can you say about the distribution of charge on each side of the membrane?

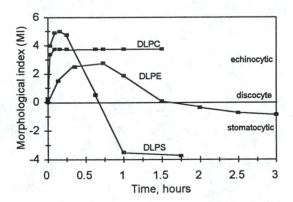

Figure 13.4: Shape changes induced by incubating red blood cells with phospholipid vesicles. (Based on Daleke and Heustis, 1985.)

2. The ability of the flippase to translocate DLPS and lyso-PS was compared. The results are shown in Figure 13.5. What does this tell you about the structural requirement for translocation activity?

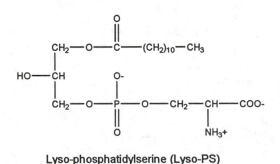

Lyso-phosphatidylserine (Lyso-PS)

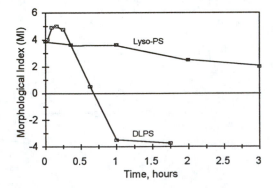

Figure 13.5: Shape changes in red blood cells induced by lyso-PS. (Based on Daleke and Heustis, 1985.)

3. Additional experiments were carried out to ascertain the involvement of ATP. Previous experiments have shown that incubating red blood cells with iodoacetamide and inosine causes ATP levels to decrease by 60%. Experiments similar to those diagramed in Figure 13.1 were carried out, but one batch of red cells was pre-treated with iodoacetamide and inosine for 30 minutes first, then phospholipid vesicles containing purified dimyristoylphosphatidylserine (DMPS) were added. The control batch of red blood cells was not treated with iodoacetamide and inosine. The results are shown in Figure 13.6. What is your interpretation of these results?

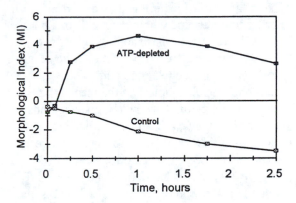

Figure 13.6: Effect of ATP depletion on red blood cell shape change experiments. (Based on Daleke and Heustis, 1985.)

4. Experiments were carried out in order to determine which amino acid side chains in the flippase enzyme were essential to its translocation ability. Red blood cells were pretreated with diamide, a reagent that modifies sulfhydryl groups as shown in Figure 13.7. After the pretreatment, the shape change experiments were carried out in the usual manner. The results are shown in Figure 13.8. What is your interpretation of these results?

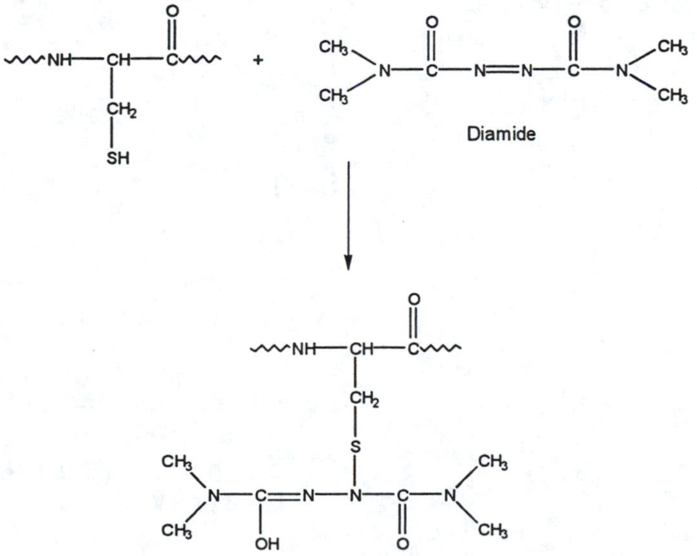

Figure 13.7: Reaction of sulfhydryl groups with diamide.

41

5. Next, a series of experiments was carried out in order to determine whether specific metal ions are required for flippase activity. First, red blood cells were treated with a mixture of an ionophore and a chelating agent. The ionophore "pokes holes" in the membrane, so that ions leak out. The chelating agent binds the ions so they can't go back inside the cell. Following this treatment, the shape change experiments were carried out as described in Figure 13.1. After 1.5 hours, Mg^{2+} ions were added back to the red blood cells. The results are shown in Figure 13.9. What is your interpretation of these results?

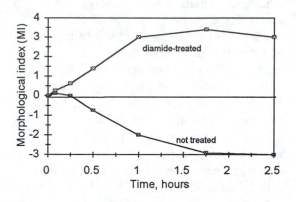

Figure 13.8: Effect of diamide treatment on DMPS-induced red blood cell shape change. (Based on Daleke and Heustis, 1985.)

6. The experiments such as the ones described here give a good picture of the characteristics of an enzyme that has yet to be completely purified. Write a paragraph that summarizes the characteristics of the flippase enzyme. Include in your description an explanation of how the flippase functions to maintain phospholipid asymmetry.

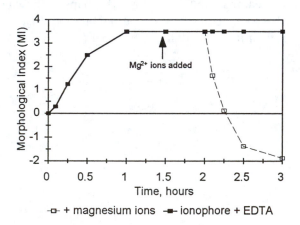

Figure 13.9: Effect of ion depletion on red blood cell shape change experiments. (Based on Daleke and Heustis, 1985.)

Reference

Daleke, D. L., and Heustis, W. H. (1985) *Biochemistry* **24**, pp. 5406-5416.

Case 14
Shavings from the Carpenter's Bench:
The Biological Role of the Insulin C-peptide

Focus concept

Recent experiments indicate that the insulin C-peptide, which is removed upon conversion of proinsulin to insulin, may have biological activity in its own right.

Prerequisites

- Amino acid structure.
- Principles of protein folding.
- Membrane transport proteins.

Background

The protein hormone insulin is synthesized as a prohormone and is processed by cleavage of two peptide bonds to yield the mature hormone, which consists of two separate polypeptide chains, A, and B, linked by disulfide bonds. The spliced out portion is referred to as the C peptide and, although the excised C peptide is secreted along with the mature insulin, scientists believed that the C peptide was biologically inactive. However, recent evidence indicates that the C peptide may have biological activity in its own right. This discovery has important implications for the treatment of diabetes.

Pharmacological doses of C-peptide administered to diabetic rats increased the level of activity of the $Na^+K^+ATPase$ pump, improved nerve conduction, vascular permeability and blood flow, but did not decrease hyperglycemia, nor did it prevent the glycosylation of many proteins seen in diabetics. Interestingly, a synthetic C-peptide made up of D-amino acids was able to bring about these same effects. Rat and human C-peptides showed the biological effects just described while C-peptides from pig and cow did not.

The authors proposed that the C-peptides were able to alter biological activity of cells by becoming integrated into the cell membrane, forming an ion channel and altering cell function. However, other researchers think that this hypothesis is unlikely.

Table 14.1: Sequences of human and animal insulins. The amino acid residues that make up the C peptide are underlined.

Rat

	1	2	3	4	5	6	7	8	9	10	11	12	13	14	15
1	Phe	Val	Asn	Gln	His	Leu	Cys	Gly	Ser	His	Leu	Val	Glu	Ala	Leu
16	Tyr	Ile	Leu	Val	Cys	Gly	Glu	Arg	Gly	Phe	Phe	Tyr	Thr	Pro	Met
31	Ser	Arg	Arg	Glu	Val	Glu	Asp	Pro	Gln	Val	Gly	Gln	Val	Glu	Leu
46	Gly	Ala	Gly	Pro	Gly	Ala	Gly	Ser	Glu	Gln	Thr	Leu	Ala	Leu	Glu
61	Val	Ala	Arg	Gln	Ala	Arg	Ile	Val	Gln	Gln	Cys	Thr	Ser	Gly	Ile
76	Cys	Ser	Leu	Tyr	Gln	Glu	Asn	Tyr	Cys	Asn					

Human

	1	2	3	4	5	6	7	8	9	10	11	12	13	14	15
1	Phe	Val	Asn	Gln	His	Leu	Cys	Gly	Ser	His	Leu	Val	Glu	Ala	Leu
16	Tyr	Leu	Val	Cys	Gly	Glu	Arg	Gly	Phe	Phe	Tyr	Thr	Pro	Lys	Thr
31	Arg	Arg	Glu	Ala	Glu	Asp	Leu	Gln	Val	Gly	Gln	Val	Glu	Leu	Gly
46	Gly	Gly	Pro	Gly	Ala	Gly	Ser	Leu	Gln	Pro	Leu	Ala	Leu	Glu	Gly
61	Ser	Leu	Gln	Lys	Arg	Gly	Ile	Val	Glu	Gln	Cys	Cys	Thr	Ser	Ile
76	Cys	Ser	Leu	Tyr	Gln	Leu	Glu	Asn	Tyr	Cys	Asn				

Cow

	1	2	3	4	5	6	7	8	9	10	11	12	13	14	15
1	Phe	Val	Asn	Gln	His	Leu	Cys	Gly	Ser	His	Leu	Val	Glu	Ala	Leu
16	Tyr	Leu	Val	Cys	Gly	Glu	Arg	Gly	Phe	Phe	Tyr	Thr	Pro	Lys	Ala
31	Arg	Arg	Glu	Val	Glu	Gly	Pro	Gln	Val	Gly	Ala	Leu	Glu	Leu	Ala
46	Gly	Gly	Pro	Gly	Ala	Gly	Gly	Leu	Glu	Gly	Pro	Pro	Gln	Lys	Arg
61	Gly	Ile	Val	Glu	Gln	Cys	Cys	Ala	Ser	Val	Cys	Ser	Leu	Tyr	Gln
76	Leu	Glu	Asn	Tyr	Cys	Asn									

Pig

	1	2	3	4	5	6	7	8	9	10	11	12	13	14	15
1	Phe	Val	Asn	Gln	His	Leu	Cys	Gly	Ser	His	Leu	Val	Glu	Ala	Leu
16	Tyr	Leu	Val	Cys	Gly	Glu	Arg	Gly	Phe	Phe	Tyr	Thr	Pro	Lys	Ala
31	Arg	Arg	Glu	Ala	Glu	Asn	Pro	Gln	Ala	Gly	Ala	Val	Glu	Leu	Gly
46	Gly	Gly	Leu	Gly	Gly	Leu	Gln	Ala	Leu	Ala	Leu	Glu	Gly	Pro	Pro
61	Gln	Lys	Arg	Gly	Ile	Val	Glu	Gln	Cys	Cys	Thr	Ser	Ile	Cys	Ser
76	Leu	Tyr	Gln	Leu	Glu	Asn	Tyr	Cys	Asn						

Questions

1. What are the physiological symptoms of untreated diabetes?

2. What is the structural purpose of the C peptide? In other words, why is insulin synthesized as a prohormone rather than being synthesized as separate A and B chains?

3. Prohormone convertase enzymes convert proinsulin to insulin. What do you note about the substrate specificity of the convertases?

4. In the human disease familial hyperinsulinemia, the arginine at position 89 has been mutated to a histidine. What are the biochemical consequences of this mutation?

5. Why do you think that a synthetic C-peptide consisting of D-amino acids (rather than the naturally occurring L amino acids) is as biologically active as the natural C-peptide which consists of L-amino acids? What can you conclude about the structural requirements for C-peptide biological activity?

6. Why do you think that the rat and human C-peptides had biological activity whereas the porcine and bovine C-peptides did not?

7. Evaluate the hypothesis that the C-peptide functions as a channel-former.

8. How is diabetes currently treated in this country? Based on the information presented here, would you modify the protocol in any way?

References

Ido, Y., Vindigni, A., Chang, K., Stramm, L., Chance, R., Heath, W. F., DiMarchi, R. D., DiCera, E., Williamson, J. R.(1997) *Science* **277**, pp. 563-566.

Steiner, D. F., and Rubenstein, A. H. (1997) *Science* **277**, pp. 531-532.

Wahren, J., Johansson, B.-L., and Wallberg-Henriksson, H. (1994) *Diabetologica [Suppl 2]* **37**, pp. S99-S107.

Case 15
Site-Directed Mutagenesis of Creatine Kinase

Focus concept

Site-directed mutagenesis is used to create mutant proteins so that the role of a single reactive cysteine in binding and catalysis can be assessed for the enzyme creatine kinase.

Prerequisites

- Amino acid structure.
- Protein architecture.
- Enzyme kinetics and inhibition.
- Basic enzyme mechanisms.

Background

The enzyme creatine kinase is important in energy metabolism and catalyzes the reaction shown in Figure 15.1. The reaction is readily reversible *in vitro*. Enzyme activity can be assayed by measuring hydrogen ion release (as determined by pH) in the forward reaction or hydrogen ion consumption in the reverse reaction.

In order to understand a particular enzyme's mechanism, scientists rely on a variety of techniques. One preferred technique is x-ray crystallography, which gives a three-dimensional picture of what the protein molecule looks like. X-ray crystallographic data are not always available however, since it is difficult to prepare the protein crystals that are used in carrying out the analysis. In the absence of crystallographic data, scientists use other experimental methods to determine the important features of an enzyme's catalytic mechanism. One technique is chemical modification in which various reagents that might derivatize specific amino acid side chains are added to an enzyme solution. Then the experimenter measures the activity of the enzyme using a specific assay, and uses sequencing techniques to determine which amino acid has been modified. If a chemical modification results in inactivation of the protein, it can be inferred that that amino acid is essential for the enzyme's activity in some way.

Another technique which is especially powerful is site-directed mutagenesis in which genetic engineering techniques are used to create mutant proteins with a single amino acid substitution. In this manner, the role of specific amino acids in the catalytic mechanism can be ascertained.

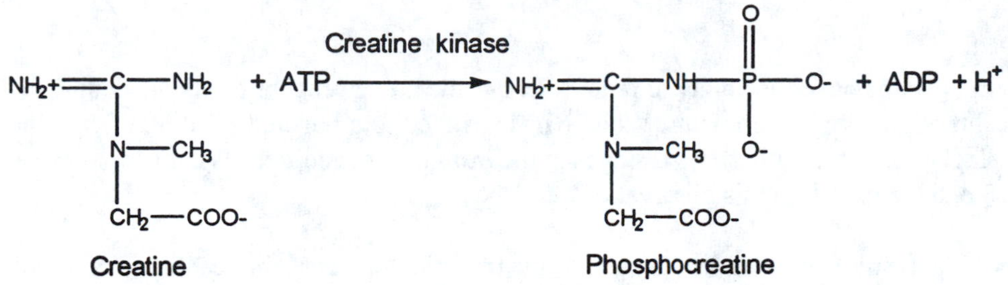

Figure 15.1: Reaction catalyzed by creatine kinase.

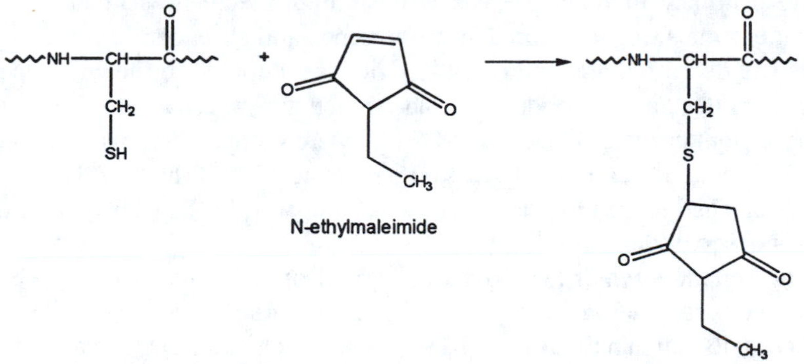

Figure 15.2: Reaction of Cys in proteins with NEM.

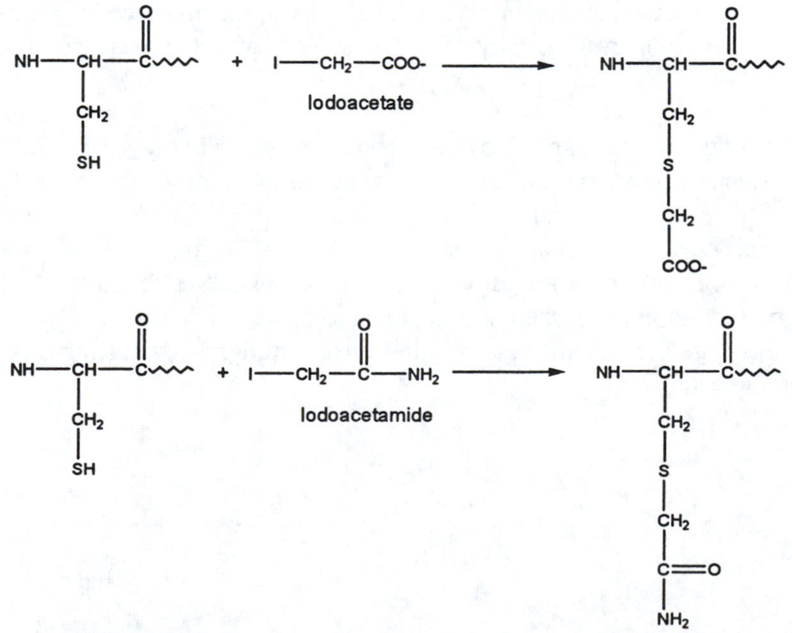

Figure 15.3: Reaction of Cys with iodoacetate and iodoacetamide.

Questions

1. Sulfhydryl groups have the ability to react with the alkylating reagent N-ethylmaleimide (NEM) in chemical modification experiments. When NEM is added to a purified solution of creatine kinase, Cys 278 is alkylated, but no other Cys residues in the protein are modified. What can you infer about the Cys 278 residue based on this observation?

2. Based on the result described in Question 1, the investigators used the technique of site-directed mutagenesis to synthesize five mutant creatine kinase proteins in which the Cys 278 was replaced with either a Gly, Ser, Ala, Asn or Asp residue. These mutants were termed C278G, C278S, C278A, C278N and C278D, respectively, to indicate the exact position of the amino acid change. The activities of the mutant enzymes were measured in the presence and absence of specific cofactors.
   a. All of the mutants had decreased creatine kinase activity as compared to the wild-type enzyme. What information does this give you about the wild-type enzyme mechanism?
   b. The activity of the mutant enzymes C278D and C278N were compared and it was found that the activity of the C278D mutant was 12-fold greater than the activity of the C278N mutant (although both mutants had lower enzyme activities as compared to the wild-type). Suggest an explanation for this observation.
   c. The activities of the mutant enzymes (although decreased from the wild-type) were enhanced when either chloride or bromide ions were added to the assay mixture. (An exception was the C278D mutant). Why do you think that the ions were able to enhance enzyme activity?
   d. The C278D mutant was an exception to the observation described in Question 2c above. This mutant did not show an enhancement of enzyme activity in the presence of chloride and bromide ions; in fact its minimal enzyme activity decreased somewhat in the presence of these ions. Explain why.

3. The results of the study presented here provided greater understanding for previous chemical modification experiments that had been carried out with creatine kinase. These reactions are shown in Figure 15.3. Some investigators modified the Cys 278 with iodoacetamide and found that the enzyme activity was abolished as a result. They concluded that the Cys 278 residue was absolutely essential for enzymatic activity. But other investigators modified the Cys 278 with iodoacetate and found that activity was decreased but not abolished, leading them to conclude that the Cys 278 was *not* essential. Taking these results together with the results of the current study, can you suggest an explanation that will clear up the confusion?

4. The investigators next carried out kinetic studies in which they measured the ability of a second substrate to bind once the first substrate had bound. The data are shown in Table 15.1. K_d refers to the constant for the binding of substrate to free enzyme and K_M refers to the constant for binding of that same substrate to the enzyme when the other substrate is already bound to the enzyme. The results for the wild type and two of the mutants are shown in Table 15.1.

   a. Compare the K_d and K_M values for creatine and ATP for the wild-type enzyme. What does a comparison of the K_d and K_M values tell you about the ability of each substrate to bind to the enzyme alone? to the enzyme when the other substrate is present?

   b. Similarly, compare the K_d and K_M values for the two mutant enzymes. Again, what does a comparison of the K_d and K_M values tell you about the ability of each substrate to bind to the enzyme alone? to the enzyme when the other substrate is present?

   c. Consider your answers to 4a and 4b and consider the V_{max} values for both the wild type and mutant enzymes. Assess the role of Cys 278 in the binding of creatine and ATP substrates to the creatine kinase enzyme.

Table 15.1: Binding constants for phosphocreatine synthesis from creatine and ATP. (Modified from Furter, *et al.*, 1993.)

Enzyme	Creatine		ATP		V_{max}, μmol/min
	K_d, mM	K_M, mM	K_d, mM	K_M, mM	
Wild-type	19.6	8.9	0.70	0.32	60.7
C278G	64	273	0.27	1.13	6.0
C278S	92	209	0.31	0.70	2.0

Reference

Furter, R., Furter-Graves, E. M., and Wallimann, T. (1993) *Biochemistry* **32**, pp. 7022-7029.

Case 16
Allosteric Regulation of ATCase

Focus concept

An enzyme involved in nucleotide synthesis is subject to regulation by a variety of combinations of nucleotides.

Prerequisites

- Properties of allosteric enzymes.
- Basic mechanisms involving regulation of metabolic pathways.

Background

Aspartate transcarbamoylase (ATCase) catalyzes an early step in the synthesis of the pyrimidine nucleotides UTP and CTP. The enzyme catalyzes the condensation of carbamoyl phosphate and aspartate to form carbamoyl aspartate. The reaction pathway is shown in Figure 16.1. The enzyme has been fairly well characterized. It is known to consist of six regulatory subunits and six catalytic subunits.

In this case, we examine the properties of ATCase isolated from *E. coli* to illustrate some of the important regulatory properties of multi-subunit enzymes. As the first enzyme in a multi-step pathway, the ATCase reaction is a logical one to regulate the synthesis of pyrimidine nucleotides. Both purine (ATP and GTP) nucleotides and pyrimidine nucleotides are needed in roughly equal amounts as substrates for DNA synthesis in rapidly dividing cells. The regulation of the ATCase enzyme ensures a proper balance of purine and pyrimidine pools in *E. coli*. The goal in this case was to identify the cellular metabolites that serve as activators and inhibitors of ATCase.

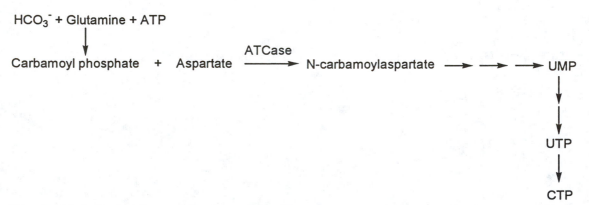

Figure 16.1: Pyrimidine synthetic pathway.

Table 16.1: Names and Abbreviations of Nucleic Acid Bases, Nucleotides and Nucleosides. (Based on Voet and Voet, 1995.)

Base formula	Base	Nucleoside	Nucleotide	Nucleotide	Nucleotide
	X = H	X = ribose	X = ribose phosphate	X = ribose tri-phosphate	X = deoxyribose tri-phosphate
	Adenine	Adenosine	AMP	ATP	dATP
	Guanine	Guanosine	GMP	GTP	dGTP
	Cytosine	Cytidine	CMP	CTP	dCTP
	Uracil	Uridine	UMP	UTP	

Questions

1. Gerhart and Pardee measured ATCase activity in the presence of a variety of purine and pyrimidine derivatives. Their results are presented in Table 16.2. What compound(s) were the most effective inhibitors? activators? Explain the significance of the metabolites that served as inhibitors or activators in the context of the biosynthetic pathway presented in Figure 16.1.

Compound	Inhibition, % (Conc = 2 mM)
Cytosine	0
Cytidine	24
CMP	38
dCMP	48
CTP	86
dCTP	88
UTP	8
GTP	35
dGTP	31
ATP	-180*
dATP	-162*

Table 16.2: Effect of nitrogen bases, nucleosides and nucleotides on ATCase activity. *Indicates stimulation. (Based on Gerhart and Pardee, 1962.)

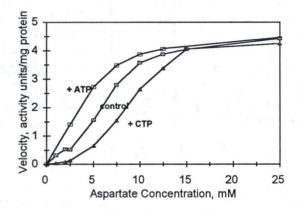

Figure 16.2: Kinetics of ATCase in the presence of ATP and CTP. (Based on Gerhart and Pardee, 1962.)

2. The kinetics of the ATCase reaction were examined using increasing concentrations of aspartate, in the presence and absence of CTP and ATP as shown in Figure 16.2.
   a. What information can you obtain by looking at the shapes of the curves in this figure?
   b. What kinetic parameter(s) change in the presence of CTP? What parameter(s) do not change? What is the significance of these observations?
   c. Answer question 2b for ATP.

3. The investigators examined the behavior of the ATCase enzyme in the presence of CTP, and in the presence of both CTP and ATP. The concentration of CTP is 0.1 mM and the concentration of ATP is 2 mM. The results are shown in Figure 16.3. What is the significance of these observations?

4. Back in 1962, Gerhart and Pardee developed a model for the regulation of the activity of the ATCase enzyme by CTP and ATP, using the pathway given in Figure 16.1. Describe that model, using information presented here as well as what you have learned about allosteric enzymes. Be sure to include a sentence explaining the physiological significance of your model.

5. Many years later, in 1989, Wild, *et al.* revisited the idea of allosteric control of ATCase by CTP. They noted that CTP did indeed inhibit ATCase, but that the inhibition was always incomplete, even at high concentrations of CTP. They hypothesized that perhaps CTP did not act alone, but in combination with some other nucleotide. They tested the activity of ATCase in the presence of several nucleotide combinations. The results are shown in Table 16.3.

a. What combination gives the most effective inhibition?

b. What is the physiological significance of this combination?

c. Revise Figure 16.1 to include this new information.

d. Redraw Figure 16.2 to include this new information. How does the K_M of the nucleotide combination compare with the values for the nucleotides alone?

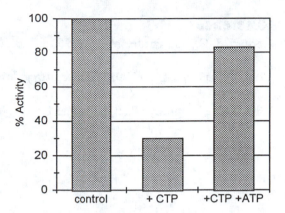

Figure 16.3: % ATCase activity in the presence of CTP, and in the presence of both CTP and ATP. (Based on Gerhart and Pardee, 1962.)

Table 16.3: Relative specific activities for combinations of nucleotide effectors. A value greater than one indicates stimulation, a value less than one indicates inhibition.

Nucleotide Effector	Aspartate Concentration	
	2.5 mM	5.0 mM
ATP	1.86	1.35
CTP	0.31	0.43
GTP	0.57	0.71
UTP	0.95	0.95
ATP/CTP	0.70	0.85
ATP/GTP	1.98	1.58
ATP/UTP	1.96	1.52
CTP/GTP	0.41	0.58
CTP/UTP	0.05	0.06
GTP/UTP	0.66	0.84

References

Gerhart, J. C., and Pardee, A. B. (1962) *J. Biol. Chem.* **237**, pp. 891-896.

Voet, D., and Voet, J. (1995) *Biochemistry*, John Wiley & Sons, New York, p. 191-194.

Wild, J. R., Loughrey-Chen, S. J., and Corder, T. S. (1989) *Proc. Natl. Acad. Sci.* **86**, pp. 46-50.

A Possible Mechanism for Blindness Associated with Diabetes: Na⁺-Dependent Glucose Uptake by Retinal Cells

Focus concept

Glucose transport into cells can influence collagen synthesis, which causes the basement membrane thickening associated with diabetic retinopathy.

Prerequisites

- Transport proteins.
- Na^+K^+ATPase and active transport.

Background

In the retina of the eye, there are equal amounts of endothelial cells and cells called pericytes. The two types of cells are associated with one another and may communicate with each other. The investigators were particularly interested in pericytes because it had been observed that pericyte damage and basement membrane thickening occur during the early stages of diabetic retinopathy. Diabetic retinopathy eventually leads to blindness.

The investigators were interested in the factors responsible for causing pericyte damage. They noted that cells in the kidney that were similar in function to the pericytes possessed a sodium-coupled glucose transporter (SGLT), and they hypothesized that the pericytes might also have such a transporter. They hypothesized that such a transporter, if it existed, would facilitate and perhaps regulate the entry of glucose into the cell. Next, the investigators determined if entry of glucose into the cell was correlated with collagen synthesis. They also investigated several inhibitors of the glucose transport process.

Questions

1. The investigators cultured pericytes and endothelial cells separately and measured their ability to take up glucose. Glucose uptake was measured both in the presence of sodium ions (provided by sodium chloride) and in the absence of sodium (choline chloride was added in order to maintain constant osmolarity). The results are shown in Figure 17.1. What is your interpretation of these results?

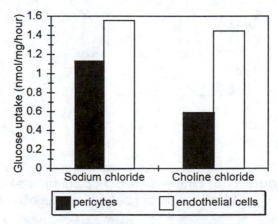

Figure 17.1: Effect of Na⁺ on glucose uptake by two different types of retinal cells. The concentrations of both NaCl and choline chloride were 145 mM. (Based on M. Wakisaka, *et al.*, 1997.)

2. Next, the investigators measured the effect of increasing sodium concentration on glucose uptake in both pericytes and endothelial cells. The results are shown in Figure 17.2. What is your interpretation of these results? What information is conveyed by the shape of the curves?

3. Transport kinetics through protein transporters can be described using the language of Michaelis-Menten. The transported substance binding to its protein transporter is analogous to the substrate binding to an enzyme. Thus, K_M and v_{max} values can be determined for protein transporters. Use the information provided in Figure 17.3 to estimate the K_M and v_{max} for glucose uptake by pericytes in the presence and absence of sodium. What information is conveyed by these values?

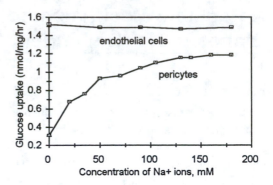

Figure 17.2: A dose-dependent response for the sodium-dependent uptake of glucose by pericytes and endothelial cells.

Have the investigators demonstrated convincingly that a SGLT exists in pericytes? What about endothelial cells? Explain, using the above data to support your answers.

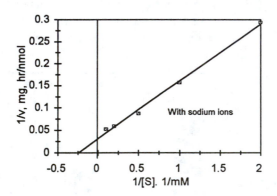
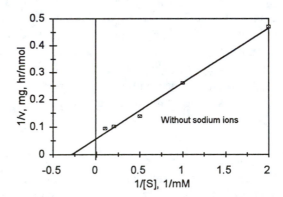

Figure 17.3: Lineweaver-Burk plots for the uptake of glucose by pericytes in the presence (145 mM NaCl) or absence (145 mM choline chloride) of sodium ions.

4. The activity of the SGLT in pericytes was investigated in the presence of inhibitors. Radioactively labeled glucose was added to cultured pericytes, in the absence (control) or presence of several potential inhibitors. The transport experiments were carried out both in the presence and absence of sodium ions. The results are shown in Figure 17.4.

   a. What is the effect of sugars such as galactose and 2-deoxyglucose on glucose transport?

   b. What is the effect of 0.2 mM phlorizin on glucose transport?

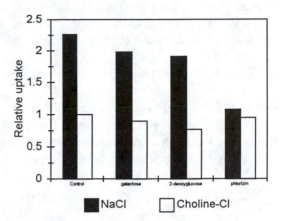

Figure 17.4: Effect of inhibitors on glucose uptake by pericytes in the presence and absence of sodium ions. The sugars were added at a concentration of 5 mM, while the phlorizin was added at a concentration of 0.2 mM.

5. Next, the investigators wanted to explore the relationship between glucose uptake and collagen synthesis. They added increasing amounts of glucose to cultured pericytes and then measured the relative density of the cells as a result. (They confirmed in a separate immunoblotting experiment that increased collagen synthesis was responsible for the increased cellular density.) The results are shown in Figure 17.5. How do you interpret these data?

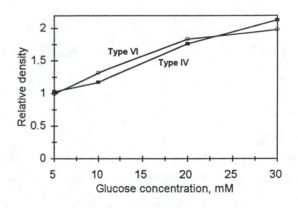

Figure 17.5: Effect of increasing glucose concentrations on the synthesis of Type IV and Type VI collagen.

6. The effect of the drug phlorizin on glucose consumption and relative cellular density (compared with 5 mM glucose) was examined. The results are shown in Figure 17.6.
   a. What effect does phlorizin have on glucose consumption at the two different concentrations of glucose?
   b. What effect does phlorizin have on relative density of the cells at the two glucose concentrations?

7. Consider the diabetic vs. the nondiabetic patient. Why do you think that retinopathy is more likely to occur in the diabetic patient? Describe the sequence of biochemical events leading to pericyte cell damage. What steps might be taken to decrease the incidence of diabetic retinopathy? Explain your answer, using the experimental results presented here.

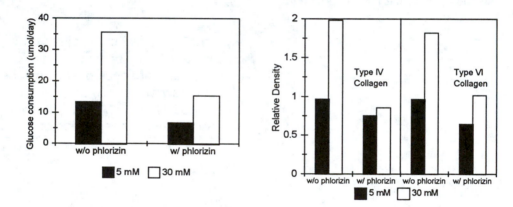

Figure 17.6: Effect of phlorizin on glucose consumption (left) and collagen synthesis (right).

Reference
Wakisaka, M., Yoshinari, M., Yamamoto, M., Nakamura, S., Asano, T., Himeno, T., Ichikawa, K., Doi, Y., and Fujishima. M. (1997) *Biochim. Biophys. Acta* **1362**, 87-96.

Case 18
Purification of Phosphofructokinase 1-C

Focus concept

The purification of the C isozyme of PFK-1 is presented and the kinetic properties of the purified enzyme are examined.

Prerequisites

- Protein purification techniques.
- Enzyme kinetics and inhibition.
- The glycolytic pathway.

Background

In this case, Foe and Kemp purified the isozyme phosphofructokinase-1 C (PFK-1 C) from brain tissue. There are three isozymes of PFK-1 and they are designated A, B, and C. The A isozyme (M_r = 84,000 dal) is found in the muscle and the brain; the B isozyme (M_r = 80,000 dal) is found in the liver and the brain; and the C isozyme (M_r = 86,000) is found in the brain. Because the brain contains all three isozymes and there isn't a location where the C isozyme is found exclusively, the enzyme has been difficult to purify. In this case, the investigators purified the desired enzyme to homogeneity, and also presented ample evidence that the C isozyme is distinct from the A and B isozymes. The availability of a pure C preparation means that antibodies can be generated which can be used to detect the isozyme. Since the levels of PFK-1 isozymes have been shown to change during malignant transformation of cells, the availability of a C antibody might be a valuable diagnostic tool.

Questions

1. To accomplish the purification, rabbit brain tissue was homogenized and centrifuged to remove insoluble material. Next, the soluble preparation was loaded on top of an ATP-Sepharose column. This is an affinity column in which ATP is covalently linked to a polysaccharide bead. The sample is loaded on top of the column, washed with a low-salt buffer, followed by a wash with a high salt buffer. What is the rationale for using this procedure? Draw a diagram of the expected elution profile.

2. Next, the fractions containing PFK-1 activity were applied to a DEAE-Sephadex (anion exchange) column. The column was equilibrated with a pH = 8.2 buffer. The column was eluted with a salt (ammonium sulfate) gradient and the results are shown in Figure 18.1. Using the elution profile as well as the results from SDS-PAGE analysis shown in Figure 18.2, identify which isozyme is found in each of the two peaks. How do you think that PFK-1 B is different from PFK-1 A and C based on the manner of elution from the DEAE column?

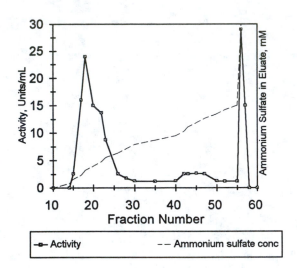

Figure 18.1: Purification of PFK-1C by DEAE Sephadex (anion exchange) chromatography. (Based on Foe and Kemp, 1985.)

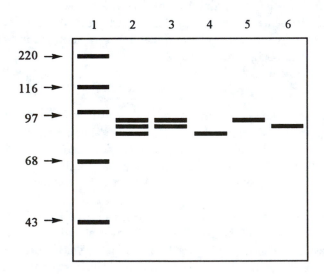

Figure 18.2: SDS-PAGE analysis of the PFK-1C purification. Lane 1: Molecular weight standards. Lane 2: Eluant from the ATP-Sepharose affinity column. Lane 3: Pooled fractions 16-26 from the DEAE-Sephadex column. Lane 4: Pooled fractions 55-60 from the DEAE-Sephadex column. Lane 5: Fractions 30-35 from the CM-52 column (elution profile shown in Figure 18.3). Lane 6: Fractions 41-50 from the CM-52 column. (Based on Foe and Kemp, 1985.)

3. Next, the investigators took Fractions 16-26 from the DEAE-Sephadex column, pooled them, and adjusted the pH to 5.0. This preparation was then loaded onto a CM-52 (cation exchange) chromatography column and eluted with a pH gradient. Fractions 30-35 were collected and pooled, as were Fractions 41-50. Identify the peaks in the chromatogram in Figure 18.3, using information in the SDS-PAGE gel. Speculate as to the structural differences between the two isozymes, based on their elution from the cation exchange column.

4. Write the reaction catalyzed by PFK-1.

5. There are several allosteric effectors that influence the activity of PFK-1 in the cell. What are they? List both activators and inhibitors of the enzyme.

6. The investigators next carried out kinetic studies using their newly purified PFK-1C isozyme. They studied the catalytic behavior of the enzyme in the presence of the metabolites AMP, inorganic phosphate (P_i) and fructose-2,6-bisphosphate (F-2,6-BP). The results are shown in Figure 18.4. Additional information concerning the three isozymes' response to allosteric effectors is presented in Tables 18.1 and 18.2.

   a. Compare the ability of PFK-1C to catalyze the phosphorylation of fructose-6-phosphate in the absence of, and in the presence of AMP, F-2,6-BP or P_i . By what mechanism do these allosteric effectors influence the velocity of the reaction?

   b. Evaluate whether the investigators have shown that PFK-1 C is different from the PFK-1 A and PFK-1 B isozymes. Speculate why there might be differences among the isozymes.

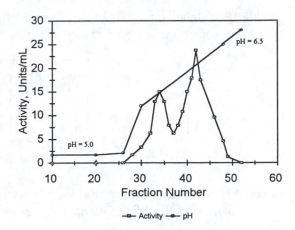

Figure 18.3: Cation exchange chromatography of PFK-1. (Based on Foe and Kemp, 1985.)

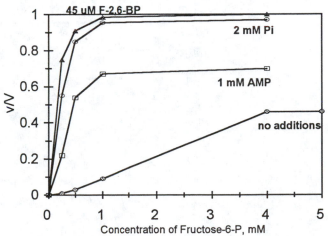

Figure 18.4: Activation of PFK-1C. V is V_{max} and is defined as the activity at 0.5 mM ATP, 10 mM fructose-6-phosphate and 2 mM P_i. (Based on Foe and Kemp, 1985.)

61

Table 18.1: Relative potency of the allosteric effector citrate on PFK isozymes. The concentration given is the micromolar concentration of citrate required to inhibit 50% of the enzyme activity.

Isozyme	Citrate, μM
A	100
B	> 2000
C	750

Table 18.2: Relative potency of allosteric effectors on PFK isozymes. Numbers given are the micromolar concentrations of each effector required to achieve 50% of the maximal velocity.

Isozyme	Phosphate	AMP	Fructose-2,6-BP
A	80 μM	10 μM	0.05 μM
B	200 μM	10 μM	0.05 μM
C	350 μM	75 μM	4.5 μM

Reference

Foe, L. G., and Kemp, R. G. (1985) *J. Biol. Chem.* **260** pp. 726-730.

Case 19
Purification of Rat Kidney Sphingosine Kinase

Focus concept

The purification and kinetic analysis of an enzyme that produces a product important in cell survival is the focus of this study.

Prerequisites

- Protein purification techniques and protein analytical methods.
- Basic enzyme kinetics.

Background

Sphingolipid metabolites have recently been shown to be involved in cell viability. For example, ceramide has been shown to stimulate apoptosis, or programmed cell death, whereas sphingosine-1-phosphate (SPP) promotes cell survival. A number of different factors influence the levels of ceramide and SPP in the cell, as shown in Figure 19.1.

Clearly, the balance of ceramide and SPP will be one of the many factors that influences whether a cell lives or dies. An understanding of the biochemical processes involved in regulating these sphingolipids has important applications in the treatment of cancer. We will concentrate on sphingosine-1-phosphate (SPP).

SPP is synthesized in the cell from sphingosine and ATP as shown in Figure 19.2. The enzyme that catalyzes the reaction is sphingosine kinase. In this case, we will study the purification of the sphingosine kinase enzyme.

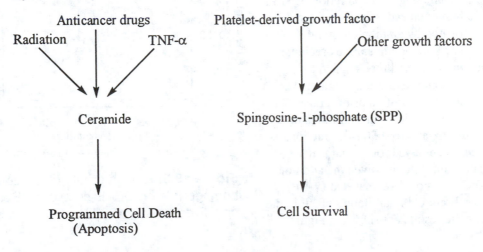

Figure 19.1: Factors that influence cellular levels of ceramide and sphingosine-1-phosphate.

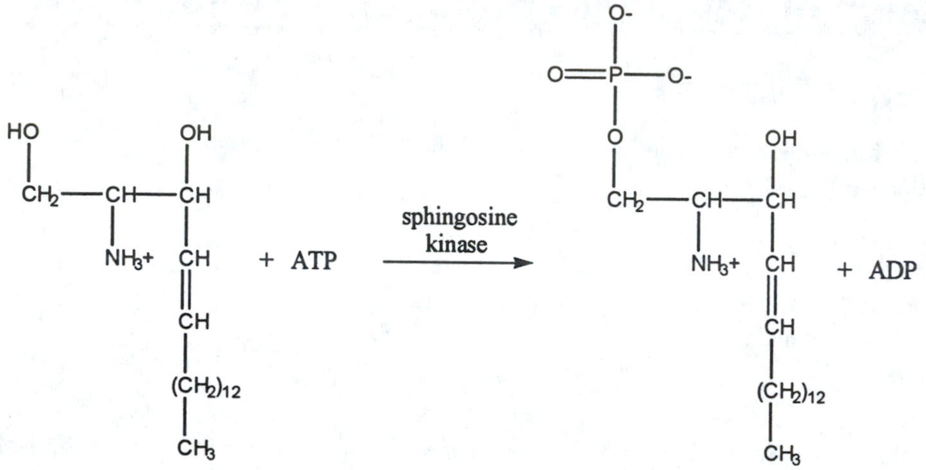

Figure 19.2: Synthesis of sphingosine-1-phosphate from sphingosine and ATP.

Questions

1. The investigators purified the enzyme sphingosine kinase using rat kidneys as a source of enzyme since it had been shown previously that kidneys contain a high concentration of SPP. Kidneys were homogenized in buffer, filtered, and then fractionated by ammonium sulfate precipitation. The ammonium sulfate precipitate was dissolved in buffer and then applied to a DEAE-cellulose (anion exchange) column. The investigators eluted the protein using Tris buffer at pH = 7.4 with increasing concentrations of sodium chloride (from 0 to 0.50 M). Fractions were collected and assayed for sphingosine kinase in a procedure that involved the addition of sphingosine to radioactively labeled (γ-^{32}P) ATP and the subsequent quantitation of ^{32}P in the SPP product. The results are shown in Figure 19.3.

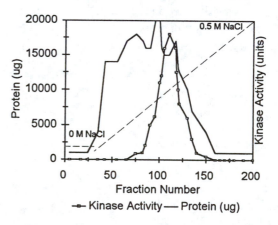

Figure 19.3: Elution profile of rat kidney homogenate on DEAE-cellulose. (Based on Olivera, *et al.,* 1998.)

Compared to the other proteins in the rat kidney, what statements can you make about the sphingosine kinase enzyme, based on its elution behavior from the DEAE column?

2. The investigators reported that the sphingosine kinase enzyme bound to the DEAE column when Tris was used as a buffer, but not when a phosphate buffer was used. The structure of the conjugate base of Tris buffer is shown in Figure 19.4. (Both buffers were adjusted to an identical pH of 7.4). Explain why.

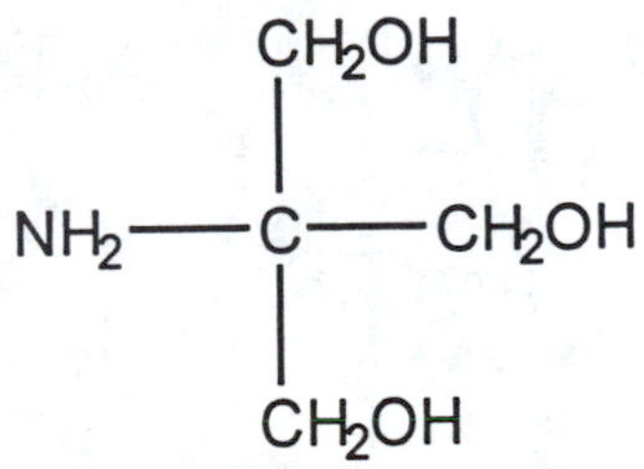

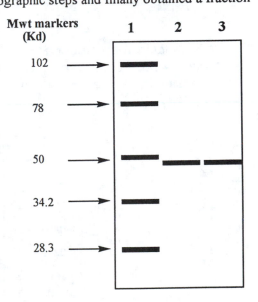

Figure 19.4: Structure of Tris conjugate base.

3. The investigators carried out several additional chromatographic steps and finally obtained a fraction that contained pure sphingosine kinase, as determined by SDS-PAGE. The molecular weight was determined by comparing the migration of the enzyme with the migration of standard molecular weight markers in Lane 1. The investigators also carried out an additional experiment in which one batch of the enzyme was treated with the reducing agent β-mercaptoethanol (Lane 2). A second batch was not treated (Lane 3). The results are shown in Figure 19.5.
   a. Estimate the molecular weight of sphingosine kinase.
   b. Interpret the results of the experiment described above. What can you say about the structure of sphingosine kinase?

Figure 19.5: SDS-PAGE gel of purified sphingosine kinase.

4. The investigators next carried out simple kinetic analyses using sphingosine and ATP as substrates in separate experiments. Michaelis-Menten plots and the corresponding Lineweaver-Burk plots are shown in Figures 19.6 and 19.7. Estimate K_M and v_{max} values for both substrates.

5. Next, the activity of sphingosine kinase was measured in the presence of *threo*-sphingosine, a stereoisomer of sphingosine. The Lineweaver-Burk plot obtained from a kinetic analysis of sphingosine kinase activity carried out in the presence of this inhibitor is shown in Figure 19.8. Estimate the K_M and v_{max} values in the presence of the inhibitor. What kind of an inhibitor is *threo*-sphingosine? Explain your answer completely.

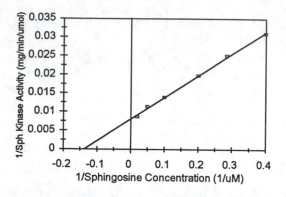

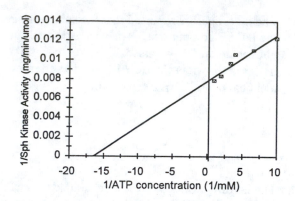

Figure 19.6: Lineweaver-Burk analysis of sphingosine kinase using varying amounts of sphingosine substrate. (Based on Olivera, *et al.*, 1998.)

Figure 19.7: Lineweaver-Burk analysis of sphingosine kinase using varying amounts of ATP substrate. (Based on Olivera, *et al.*, 1998.)

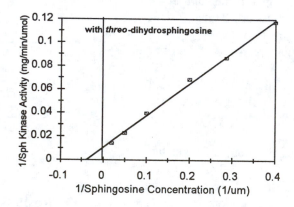

Figure 19.8: Lineweaver-Burk analysis of sphingosine kinase in the presence of the inhibitor *threo*-sphingosine. (Based on Olivera, *et al.*, 1998.)

Reference

Olivera, A., Kohama, T., Tu, Z., Milstein, S., and Spiegel, S. (1998) *J. Biol. Chem.* **273**, 12576-12583.

Case 20
NAD⁺-dependent Glyceraldehyde-3-phosphate Dehydrogenase from *Thermoproteus tenax*

Focus concept

Glycolytic enzymes from *Thermoproteus tenax* are regulated in an unusual manner.

Prerequisites

- The glycolytic pathway.
- Enzyme kinetics and inhibition.

Background

Carbohydrate metabolism in the thermophilic archaeal bacterium *Thermoproteus tenax* is rather peculiar compared to the types of organisms usually studied in introductory biochemistry. For example, the phosphofructokinase reaction in *T. tenax* is reversible, and is dependent upon pyrophosphate rather than ATP. In addition, *T. tenax* has two different glyceraldehyde-3-phosphate dehydrogenase isoenzymes. One is well known and requires NADP⁺ as a cofactor. Because of the involvement of NADP⁺, it has been hypothesized that the role of the NADP⁺-dependent enzyme is to provide NADPH for anabolic reactions and serine biosynthesis. The second isoenzyme requires NAD⁺ as a cofactor and is less well-studied, but the involvement of NAD⁺ suggests a more traditional role of the enzyme in glycolysis. In this case, we will consider the properties of the NAD⁺-dependent glyceraldehyde-3-phosphate dehydrogenase.

Although most regulatory studies of glycolysis focus on the three irreversible kinase reactions, glyceraldehyde-3-phosphate dehydrogenase (GAPDH) also has interesting regulatory properties. Glycolysis is an important pathway in the *T. tenax* organism, as it is the pathway via which most of the cell's energy is derived. *T. tenax* stores energy in the form of glycogen, which is degraded via glycogen-olysis. The products of glycogenolysis then enter the glycolytic pathway.

The gene for GAPDH was purified and its kinetic characteristics studied. Summary information is presented in Table 20.1.

Table 20.1: Kinetic properties of NAD⁺-dependent GAPDH isolated from *T. tenax.* (Based on Brunner, *et al.,* 1998.)

NAD⁺ saturation	
Without AMP	
V_{max}, units/mg	36.5
K_M, mM	3.3
With AMP	
V_{max}, units/mg	37.0
K_M, mM	1.4
Thermal stability	
Residual activity after 100 min at 100°C	
	30 %
Molecular Mass	
Subunit (kD)	55,000
Native (kD)	220,000

Questions

1. Name the three enzymes that catalyze irreversible, regulated reactions in glycolysis.

2. Write the balanced equation of the reaction catalyzed by NAD⁺-dependent glyceraldehyde-3-phosphate dehydrogenase. Include structures and cofactors.

3. What is the importance of the GAPDH reaction to glycolysis?

4. The GAPDH enzyme from *T. tenax* did not show any activity in the presence of 1,3-bisphosphoglycerate. What is the significance of this observation?

5. The activity of the GAPDH enzyme was assayed in the presence of a constant amount of glyceraldehyde-3-phosphate and an increasing amount of NAD⁺. The activity of the control was compared to the activity in the presence of various metabolites. The results are shown in Figure 20.1. Additional data is given in Table 20.2.
   a. Classify the metabolites listed in Table 20.1 as inhibitors or activators. Explain your reasoning.
   b. What is the physiological significance of your answer to Question 5a?

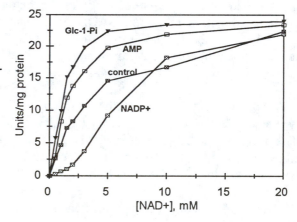

Figure 20.1: The activity of NAD⁺-dependent GAPDH activity in the presence of various effectors. (Modified from Brunner, *et al.*, 1998.)

Table 20.2: Effect of various metabolites on the activity of NAD^+-dependent GAPDH isolated from *T. tenax*. (Based on Brunner, *et al.*, 1998.)

Metabolite	Apparent K_M, mM
None	3.1
$NADP^+$	4.5
Glucose-1-phosphate	0.4
NADH	8.0
AMP	1.3
ADP	1.7
ATP	30

6. What does the information in Figure 20.1 (and other information available to you) tell you about the mechanism of inhibition and activation of GAPDH by these selected metabolites?

Reference

Brunner, N. A., Brinkmann, H., Siebers, B., and Hensel, R. (1998) *J. Biol. Chem.* **273**, pp. 6149-6156.

Case 21
Characterization of Pyruvate Carboxylase from *Methanobacterium thermoautotrophicum*

Focus concept

Pyruvate carboxylase is discovered in a bacterium that was previously thought not to contain the enzyme.

Prerequisites

- Tricarboxylic acid cycle reactions and associated anapleurotic reactions.
- Glyoxylate cycle reactions.

Background

As one of the reactants in the first reaction of the tricarboxylic acid cycle, oxaloacetate (OAA) is an important cellular metabolite. The concentrations of oxaloacetate are tightly regulated.

Different organisms employ different mechanisms to obtain oxaloacetate. In mammals and yeast, OAA is the product of the pyruvate carboxylase (PYC) reaction. In *E. coli*, the enzyme phosphoenol-pyruvate carboxylase (PPC) provides oxaloacetate from phosphoenolpyruvate obtained from glucose oxidation in the glycolytic pathway. If glucose is absent and *E. coli* is using acetate as a carbon source, the glyoxylate pathway serves to generate the needed oxaloacetate. Usually an organism will employ PPC or PYC, but not both.

Detectable levels of PYC in the methanogenic bacterium *Methanobacterium thermoautotrophicum* had previously not been found, and since PPC had been detected, it was believed that *M. thermoauto-trophicum* did not possess PYC. However, in the case described here, microbiologists found that if they added biotin to cultures of the methanogenic bacterium, pyruvate carboxylase activity could be detected. This was a surprising finding, especially since it is known that the methanogen can synthesize its own biotin. However, having identified the presence of the PYC enzyme, the investigators set out to isolate, purify, and characterize the enzyme. Purification of the PYC was rather straightforward since the enzyme is soluble and hydrophilic. In addition, the investigators were able to make use of the protein avidin, which binds to biotin with high affinity and specificity.

Questions

1. Write the balanced reaction catalyzed by pyruvate carboxylase (PYC), including structures and cofactors.

2. Write the balanced reaction catalyzed by phosphoenolpyruvate carboxylase (PPC), including structures.

3. Draw a diagram that describes how *E. coli* obtains oxaloacetate via the glyoxylate cycle when acetate, not glucose, is provided as a food source.

4. Why is it so important that oxaloacetate be generated in the cell?

5. Why did the investigators add biotin to the methanogen cultures in an attempt to detect pyruvate carboxylase activity?

6. Following purification of the PYC enzyme by avidin-Sepharose affinity chromatography, the investigators carried out several experiments to characterize the enzyme. First they ran samples of the enzyme on denaturing and non-denaturing gels. The results are shown in Figure 21.1. In addition, they ran the protein through a calibrated gel filtration column. The results from the gel filtration column indicated that the PYC enzyme had a molecular weight of 540 kilodaltons. Use this information to determine the structure of the PYC enzyme.

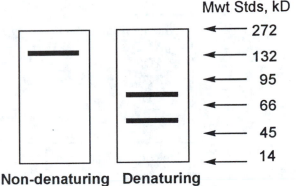

Non-denaturing   Denaturing

Figure 21.1: Gel electrophoresis of pyruvate carboxylase purified from *M. thermo-autotrophicum.* (Based on Mukhopadhyay, *et al.*, 1998.)

7. The catalytic properties of the pyruvate carboxylase enzyme were assessed following purification. The activity of the enzyme was assayed in the presence of ATP, pyruvate, bicarbonate, and Mg^{2+} ions as a control. In addition, the dependence of the enzyme on these various metabolites was tested by replacing them with similar compounds. The activity of the enzyme in the presence of these various effectors is shown in Table 21.1.
   a. Explain why PYC had no activity in the presence of avidin.
   b. Is PYC dependent on ATP for activity? Can other nucleotides substitute for ATP? What is the effect if other nucleotides are added to the assay mixture in addition to ATP?
   c. What is the effect of other tricarboxylic acid metabolites on PYC activity in the methanogen?
   d. What ion or ions are required for PYC activity?

8. The sequence of PYC from the methanogen was compared to other pyruvate carboxylase enzymes and it was discovered that the lysine at position 534 is strictly conserved. Why is this the case?

Table 21.1: Activity of pyruvate carboxylase enzyme (PYC) isolated from *M. thermo-autotrophicum* in the presence of various effectors. (Based on Mukhopadhyay, *et al.*, 1998.)

Effector	Activity (% of control)
Control	100
Avidin	0
Alternate nucleotides replacing ATP	
AMP	0
ADP	0
CTP	0
GTP	0
ITP	0
UTP	0
Nucleotides in addition to ATP	
AMP	104
ADP	73
CTP	106
GTP	94
ITP	80
UTP	105
Tricarboxylic acid cycle-related compounds	
Acetyl-CoA	84
Aspartate	91
Glutamate	95
α-ketoglutarate	73
Divalent cations replacing Mg^{2+}	
Mn^{2+}	17
Co^{2+}	46
Zn^{2+}	0

Reference

Mukhopadhyay, B., Stoddard, S. F., and Wolfe, R. S. (1998) *J. Biol. Chem.* **273**, pp. 5155-5166.

Case 22
Carrier-mediated Uptake of Lactate in Rat Hepatocytes

Focus concept

The structural characteristics of the lactate transport protein in hepatocytes are determined.

Prerequisites

- Transport proteins.
- Major carbohydrate metabolic pathways including glycolysis and gluconeogenesis.

Background

The goal of the investigation presented in this case was to determine the mechanism for lactate transport in the liver. Lactate is a major substrate for gluconeogenesis, but its transport from the blood to liver cells is incompletely understood. Lactate transport is physiologically important because the conversion of lactate to glucose in the liver as part of the Cori cycle requires that lactate released from the muscle during anaerobic fermentation be transported to the liver via the bloodstream. Previous studies have indicated that the transport of lactate is probably mediated by a carrier protein, but it was not understood exactly how the carrier worked. For example, it was not known whether lactate transport was linked to the transport of other ions such as H^+ or OH^-. In the experiments presented here, investigators measured the rate of uptake of $[^{14}C]$-labeled lactate in rat hepatocytes under a variety of different physiological conditions. In this manner, the characteristics of the lactate transporter were determined.

Questions

1. Transport of lactate from the blood to the liver is part of the Cori cycle. Draw a diagram of the Cori cycle. Under what physiological conditions would you expect the Cori cycle to be active?

2. Consider the structure of lactate–do you think that lactate would easily pass through the plasma membrane by simple diffusion? Explain why or why not.

3. The time course for the uptake of D-lactate and L-lactate by cultured rat hepatocytes is shown in Figure 22.1. What conclusions can you draw from looking at these curves?

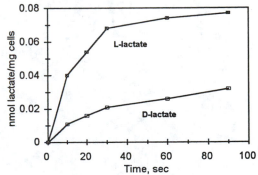

Figure 22.1: D- and L-lactate transport in rat hepatocytes. (Based on Fafournoux, *et al.*, 1985.)

4. The investigators next measured the transport of lactate in hepatocytes as a function of extracellular pH. Their results are shown in Figure 22.2. What is your interpretation of these results? Do these results give you more information about the characteristics of the lactate carrier protein?

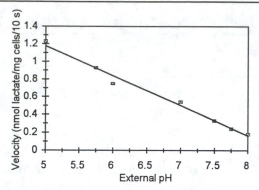

Figure 22.2: Lactate transport at various extracellular pH values. (Based on Fafournoux, *et al.*, 1985.)

5. The investigators considered the possibility that lactate did not have its own transporter, but instead used the general anionic transporter in liver cells. In order to test this hypothesis, they measured the uptake of 0.4 mM [^{14}C] lactate in the presence of 10 mM Na_2SO_4 and found that lactate transport was inhibited 11% in the presence of the Na_2SO_4. (The sulfate ion is transported by the general anionic transporter.) Do these results indicate that lactate utilizes the general anionic transporter? Explain.

Reference

Fafournoux, P., Demigné, C., and Rémésy, C. (1985) *J. Biol. Chem.* **260**, pp. 292-299.

Case 23
The Role of Uncoupling Proteins in Obesity

Focus concept

The properties of adipose tissue factors that uncouple oxidative phosphorylation are discussed and possible links between uncoupling proteins and obesity are examined.

Prerequisites

- Electron transport and oxidative phosphorylation.
- Mechanisms of uncoupling agents, such as 2,4-dinitrophenol.
- Fatty acid oxidation.

Background

Hibernating animals and human infants contain *brown fat* deposits, so-called because of the presence of large numbers of mitochondria, the site of electron transport and oxidative phosphorylation. In brown fat, given the appropriate stimulus, oxidative phosphorylation and electron transport can be uncoupled, causing energy to be dissipated as heat. The protein responsible for the uncoupling is a brown fat inner mitochondrial membrane protein previously named UCP (for uncoupling protein), but now referred to as UCP1, since a second uncoupling protein has since been discovered. Previous experiments have shown that UCP1 protects against cold and is involved in regulation of energy expenditure.

The ability of UCP1 to stimulate the consumption of calories solely for the production of heat led some investigators to postulate that UCP1 was involved in regulating body weight. Scientists have always wondered why some people seem to be able to ingest a large number of calories without gaining weight, whereas others eat moderately but are obese. If the UCP1 of brown fat were involved, scientists postulated that obese people would be efficient "burners", whereas humans of moderate weight might burn calories inefficiently, releasing a greater proportion of energy as heat. But the role of UCP1 in humans has always been debated since infants contain a large amount of brown fat but mature adults do not.

In order to examine the biochemical role of UCP1 more fully, the investigators in this case worked with mice referred to as *knockouts*. Knockout mice have been genetically engineered such that the gene coding for a particular protein is missing. By examining the characteristics of knockout mice, the biochemical and physiological roles of a particular protein can be ascertained. The investigators produced UCP1-knockout mice that are missing the gene for the UCP1 protein. They carried out experiments using these mice that are described below, studies that led to the discovery of a second uncoupling protein referred to as UCP2. The UCP2 protein may play a more significant role in obesity, since UCP2 is found in abundant amounts in white fat.

Questions

1. Why is it that heat is produced when the UCP1 protein uncouples oxidative phosphorylation from electron transport?

2. The compound 2,4-dinitrophenol (DNP) also causes uncoupling of oxidative phosphorylation and electron transport. DNP is lipid soluble and literally dissolves in the inner mitochondrial membrane. Its structure is shown in Figure 23.1. It has an acidic phenolic hydrogen which is lost in the somewhat higher pH environment of the mitochondrial matrix. The resonance stabilized anion that results is more lipid soluble than a fully charged ion would be and also has the ability to cross the inner mitochondrial membrane.
   a. Describe the mechanism of the uncoupling.
   b. DNP has been used in the past as a weight-loss aid, but it proved to be toxic and even caused a few deaths before it was taken off the market. Why might someone with a limited knowledge of biochemistry think that DNP would be an effective diet aid? Why did the compound eventually cause death in some of the people who took it?

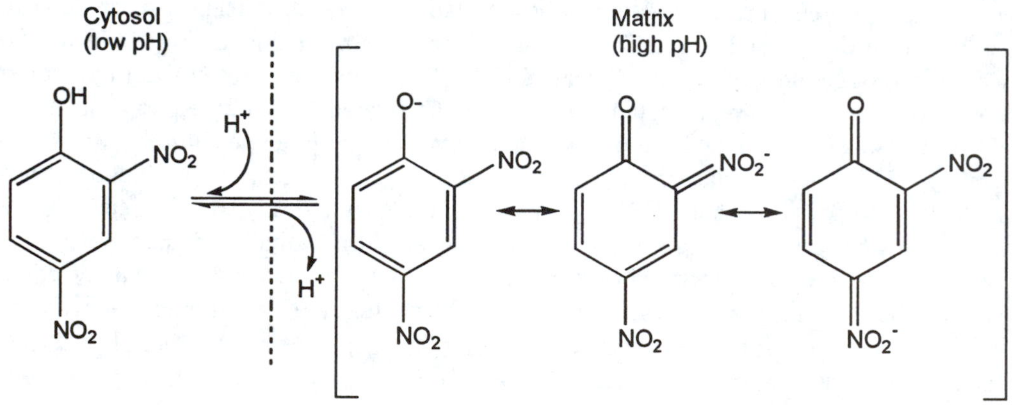

Figure 23.1: The lipid compound 2,4-dinitrophenol is also a weak acid.

3. In their study, the investigators injected a β-3 adrenergic agonist which stimulated UCP1 in adipose tissue. When they injected the agonist into normal mice, they noted that oxygen consumption increased over two-fold. But when the agonist was injected into knockout mice, oxygen consumption increased only slightly. Explain these results.

4. The investigators who produced the UCP1 knockout mice noted that the knockouts were normal in every way except that there was increased lipid deposition in their adipose tissue. Explain why.

5. The investigators carried out an experiment in which normal mice and UCP1 knockout mice were placed in a cold (5°C) room for 24 hours. The normal mice were able to maintain their body temperature at 37°C even after 24 hours in the cold. But when the knockout mice were placed in a cold room, their body temperature decreased 10°C or more. Explain.

6. Investigators wanted to know whether UCP1 synthesis could be induced by overeating, and if so, if it could be involved in burning excess fat such that an organism's adipose tissue content remained relatively constant. Along these lines, it was assumed that UCP1 knockout mice, unable to synthesize UCP1, would become obese if fed a high fat diet. Interestingly, knockout mice did not become obese even when fed a high fat diet. Instead, overfeeding caused the synthesis of a second protein to be induced. Separate experiments indicated that this second protein was also able to uncouple oxidative phosphorylation, and this protein was given the name UCP2. How would the induction of UCP2 help animals maintain a constant amount of adipose tissue?

References

Enerbäck, S., Jacobsson, A., Simpson, E. M., Guerra, C., Yamashita, H., Harper, M.-E., and Kozack, L. P. (1997) *Nature* **387**, pp. 90-94.

Fleury, C., Neverova, M., Collins, S., Raimbault, S., Champigny, O., Levi-Meyrueis, C., Bouillard, F., Seldin, M. F., Surwit, R. S., Ricquier, D., and Warden, C. H. (1997) *Nature Genetics*, **15**, pp. 269-272.

Lowell, B. B., S-Susulic, V., Hamann, A., Lawitts, J. A., Himms-Hagen, J., Boyer, B. B., Kozak, L. P., and Flier, J. S. (1993) *Nature* **366**, pp. 740-742.

Hirsch, J. (1997) *Nature* **387**, pp. 27-28.

Case 24
Uncoupling Proteins in Plants

Focus concept

Uncoupling proteins in plants uncouple oxidative phosphorylation in order to generate heat in the developing plant.

Prerequisites

- Electron transport and oxidative phosphorylation.
- Mechanisms of uncoupling agents, such as 2,4-dinitrophenol.

Background

In mammals, body temperature can be regulated by uncoupling proteins which act by uncoupling electron transport and oxidative phosphorylation. Two such uncoupling proteins have been identified which have been named UCP1 and UPC2. UCP1 is believed to be involved in heat production, whereas UCP2 might be involved in obesity in humans.

The mechanism of thermogenesis in plants is not as well understood. Several studies have been carried out which indicate that thermogenesis is an important process in the developing plant. For example, the Eastern skunk cabbage has the ability to maintain the temperature of the plant 15-35°C higher than ambient temperatures during the months of February and March where ambient temperatures range from -15 to +15°C. Thermogenesis in the skunk cabbage is critical to the survival of the plant since the spadix tissue is not frost-resistant. The skunk cabbage may be unique in its ability to carry out thermogenesis for longer than just a few hours.

Thermoregulation has also been studied in lotus flowers. Investigators in Australia noted that the temperature of the lotus flower receptacle rises one or two days before the flower opens. Temperatures fall only after the flower is completely opened. The investigators speculated that heat production in the lotus flower serves to enhance evaporation of the floral scent that attracts pollinating insects. Noting that many pollinating insects require a thoracic temperature of 30°C or higher to initiate flight, the investigators hypothesized that heat production in the lotus flower would assist the departure of the insects from the flower.

Although observations of temperature and oxygen utilization in the skunk cabbage and lotus flower implicated an uncoupler, an uncoupling protein was not isolated until fairly recently. A research team in France recently reported that they had identified an uncoupling protein in potatoes (*Solanum tuberosum*) that is involved in thermoregulation in plants in a manner similar to the *brown fat* uncoupling proteins in mammals. The uncoupling protein is named StUCP and its synthesis is induced by cold temperatures. The authors note that heat production concomitant with a "burst of respiration" occurs during plant flowering and fruit ripening.

Questions

1. Explain how uncoupling agents work and how an uncoupling agent acts to bring about an increase in temperature.

2. In the skunk cabbage, the site of thermogenesis is the spadix. However, the spadix tissue does not store starch in its tissues. Instead, the spadix relies on the massive root system in the skunk cabbage which stores appreciable quantities of starch. The large quantity of starch available may explain why the skunk cabbage is able to carry out thermogenesis in weeks rather than hours. Why is starch required for thermogenesis?

3. Oxygen consumption by the skunk cabbage increases as temperature decreases. The rate of oxygen consumption nearly doubles with every 10°C drop in ambient temperature. Similarly, in the lotus flower, the receptacle temperature is maintained between 30 and 35°C for a two day period during flower opening. The ambient air temperature during this two-day period may fluctuate between 10 and 30°C. Oxygen consumption was observed to decrease during the day when temperatures were close to 30°C. Explain the biochemical mechanism for these observations.

4. In the skunk cabbage, lotus flower, and the potato, thermogenesis occurs because of the action of an uncoupling protein. The gene that codes for the uncoupling protein in potatoes was recently isolated. The results from one of the experiments is shown in Figure 24.1. What is your interpretation of these results? How is the production of the uncoupling protein regulated?

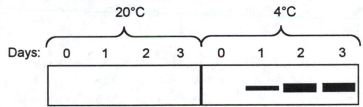

Figure 24.1: Northern blot analysis of the uncoupling protein from potatoes at 20°C and 4°C. Northern blots detect levels of mRNA. (Based on Laloi, *et al.*, 1997.)

References

Laloi, M., Klein, M., Riesmeier, Müller-Röber, B., Fleury, C, Bouillaud and Ricquier, D. (1997) *Nature* **389**, pp. 135-136.

Knutson, R. M. (1974) *Science*, **186**, pp. 746-747.

Seymour, R. S., and Schultze-Motel, P. (1996) *Nature*, **383**, p. 305.

Case 25
Glycogen Storage Diseases

Focus concept

Disturbances in glycogen utilization result if an enzyme involved in glycogen synthesis or degradation is missing.

Prerequisites

- Glycogen synthesis and degradation pathways.
- The link between glycogen metabolism and fat metabolism.

Background

Glycogen storage diseases are so named because a hallmark of the diseases is impaired glycogen storage due to a deficiency of one of the enzymes involved in either glycogen synthesis or glycogen degradation. Eight kinds of glycogen storage diseases have been identified so far. Each type is characterized by the lack of a specific enzyme. An understanding of glycogen metabolism is essential in the proper treatment of this disease, and identification of the deficient enzyme is required before a treatment protocol can be designed.

Your patient is a fifteen-year-old Caucasian male named Alex K. Alex's mother has brought him to see you because she is concerned about his inability to perform any kind of strenuous exercise. During his physical education classes, Alex could not keep up with his classmates and often suffered painful muscle cramps if he did attempt to exercise. He appeared to be normal if at rest or performing light to moderate exercise. A physical examination reveals that his liver appears to be normal in size, but his muscles are flabby and poorly developed. A fasting glucose test showed that Alex was not hypo- or hyperglycemic. A number of biochemical tests were carried out to identify the type of glycogen storage disease in this patient.

Questions

1. You decide to try Alex's response to glucagon. This test consists of injecting a high dose of glucagon intravenously and then drawing samples of blood periodically and measuring the glucose content of the samples. After the glucagon injection, Alex's blood sugar rises dramatically. Is this the response you would expect in a normal person? Explain.

2. Liver and muscle biopsies are taken from Alex and analyzed. The biopsies reveal that glycogen content in the liver is normal, but muscle glycogen content is elevated. The biochemical structure of glycogen in both tissues appears to be normal. Suggest some possible explanations for these observations.

3. Next, you do another test where you have Alex perform ischemic exercise for as long as he is able to do so. Blood is withdrawn from the patient every few minutes or so during the exercise period.

    a. Alex's blood samples are tested for lactate and compared with a control sample of a patient who does not suffer from a glycogen storage disease. The results are shown in Figure 25.1. Why does lactate concentration increase in the normal patient? Why is there no corresponding increase in Alex's lactate concentration?

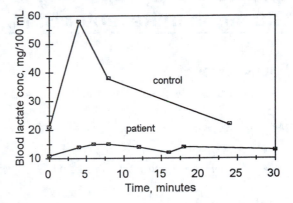

    b. Urine tests after Alex has completed his exercise reveal the presence of myoglobin in his urine. Myoglobin isn't normally found in the urine, but in muscle cells. Why does Alex suffer from myoglobinuria following ischemic exercise?

Figure 25.1: Blood lactate concentration following ischemic exercise in a patient with a glycogen storage disease and a control. (Based on Stanbury, *et al.*, 1978.)

    c. Alex's blood samples are also tested for alanine content. In a normal person, you would expect that blood alanine would increase during ischemic exercise. But in Alex's blood samples, you see a decrease in alanine concentrations, leading you to believe that Alex's muscle cells are taking up alanine rather than releasing it. Why would blood alanine concentrations increase in a normal person? Why do blood alanine concentrations decrease in your patient?

4. Alex's enzyme deficiency does not cause him to suffer from either hypo- or hyperglycemia. Explain why.

5. As a treatment, you tell Alex that the best thing he can do is to avoid strenuous exercise. If he does wish to exercise, you advise him to consume sports drinks containing glucose or fructose frequently while exercising. Why would this help alleviate Alex's suffering during exercise?

Reference

Stanbury, J. B., Wyngaarden, J. B., and Fredrickson, D. S. (1978) *The Metabolic Basis of Inherited Disease*, McGraw-Hill Book Company, New York, pp. 151-153.

Case 26
The Role of Specific Amino Acids in the Peptide Hormone Glucagon in Receptor Binding and Signal Transduction

Focus concept

Amino acid side chains important in glucagon binding and signal transduction are identified.

Prerequisites

- Amino acid structure.
- Signal transduction via G proteins.

Background

Glucagon is a 29-amino acid peptide hormone secreted by the pancreatic α-cells in response to low glucose concentrations. Its primary amino acid sequence is shown in Table 26.1. Glucagon acts primarily on the liver where binding to specific extracellular receptors stimulates glycogenolysis and gluconeogenesis with subsequent release of glucose from the liver for the benefit of other body tissues. Glucagon is counter-regulatory to insulin which is secreted by pancreatic β-cells and stimulates cellular uptake of exogenous glucose from the blood. During feeding insulin levels are high and glucagon levels are low. The opposite is true during fasting–glucagon levels rise and insulin concentrations decrease.

The glucagon hormone has been the subject of much research interest in decades past, not just because of its importance in carbohydrate metabolism, but also because its mechanism of action, via the activation of a G-protein-linked enzyme, is a model for signal transduction. But more recently attention has focused on the role of glucagon in the disease diabetes mellitus. Several studies have shown that in diabetics the lack of insulin is accompanied by hypersecretion of glucagon. The excess glucagon secretion leads to release of glucose from the liver, which exacerbates the high blood glucose concentrations in the untreated diabetic.

Diabetics are currently treated with exogenous insulin. But some investigators have suggested that the treatment regimen of the diabetic should address the glucagon hypersecretion as well as the lack of insulin. One way to do this would be to administer a glucagon antagonist along with insulin. A glucagon antagonist is a molecule that would be able to bind to extracellular receptors on liver cells, but would be unable to carry out the signal transduction process. The glucagon antagonist would compete for binding with endogenous glucagon. If the antagonist bound instead of the endogenous glucagon, glycogenolysis would not occur.

In order to construct a glucagon antagonist it is necessary to determine exactly which amino acids contribute to receptor binding and which amino acids are involved in signal transduction. These experiments were first carried out in the mid-seventies, but recent advances in biotechnology have facilitated the process. For example, the glucagon receptor gene has been cloned and sequenced, and studies have shown that an aspartate residue near the C-terminus of the receptor protein is essential for glucagon binding.

Retaining the amino acid residues important for binding while modifying those amino acids involved in signal transduction would result in a glucagon antagonist. Many such compounds have been synthesized, but the search for the ideal antagonist has been complicated by the fact that several amino acid residues in the glucagon molecule have been found to be important for both receptor binding and signal transduction.

In the current study, the investigators used the technique of solid phase peptide synthesis to synthesize modified glucagon molecules. They carried out two separate studies. The first study examined the role of amino acid residues at positions 9, 15, and 21. The second study examined the role of amino acid residues at positions 1,12, 17, and 18. In each study, amino acid residues were replaced with amino acids with different properties, and the resulting analogs were tested for their ability to bind to liver membrane receptors and carry out signal transduction. A true antagonist would be able to bind to receptors while eliciting no response whatsoever. Analogs capable of binding with diminished (but not abolished) activity are referred to as partial agonists.

Table 26.1: Primary sequence of human glucagon.

	1	2	3	4	5	6	7	8	9	10	11	12	13	14	15
1	His	Ser	Gln	Gly	Thr	Phe	Thr	Ser	Asp	Tyr	Ser	Lys	Tyr	Leu	Asp
16	Ser	Arg	Arg	Ala	Gln	Asp	Phe	Val	Gln	Trp	Leu	Met	Asn	Thr	

Questions

1. Glucagon carries out its biological function by binding to extracellular hepatic receptors and then putting into motion a series of events which leads to glycogenolysis. Draw a diagram which describes the steps of this process.

2. Why did the investigators choose amino acids at positions 1, 12, 17, and 18 for modification?

3. The investigators synthesized a number of glucagon analogs which are listed in Table 26.2. The ability of the glucagon analogs to bind to receptors and elicit a biological response was measured and compared to native glucagon. Use the information provided in the table to answer the following questions.
   a. What is the effect of substituting or eliminating the amino acid at position 9?
   b. What is the effect of the amino acid replacement or modification at position 12?
   c. What is the effect of the amino acid replacement at position 17? Be specific.
   d. What is the effect of the amino acid replacement at position 18? Be specific.
   e. What is the role of the histidine at position 1?

4. Write a summary paragraph describing the important findings of this study.

5.   Of the glucagon analogs presented here, which is the best glucagon antagonist? Could you design a better glucagon antagonist than the analogs presented here? Explain the rationale for your design.

Table 26.2: Glucagon analogs with various amino acid replacements. Binding affinity refers to the ability of the glucagon analog to bind to hepatic membrane receptors. Activity was measured by testing each analog's ability to stimulate cAMP production as compared to native glucagon. The *des* prefix indicates that the specified amino acid has been deleted. (Based on Unson, *et al.*, 1994 and 1998.)

Glucagon analog	Binding affinity	% of maximum activity
Glucagon	100%	100%
Des-Asp^9	45%	8.3%
Lys^9	54%	0%
N$^\epsilon$-acetyl-Lys^{12}	47	90.5
Ala^{12}	17.3	59.7
Gly^{12}	11.4	85.7
Glu^{12}	1.0	80.4
Ala^{17}	38	29
Leu^{17}	30	88
Glu^{17}	21.3	94.8
Ala^{18}	13	94.4
Leu^{18}	56	95
Glu^{18}	6.2	100
Des-His^1	63	44
Des-His^1-Des-Asp^9	7	0
Des-His^1-Lys^9	70	0
Des-His^1-Glu^{12}	0.11	28
Des-His^1-Glu^{17}	1.7	21.5
Des-His^1-Glu^{18}	0.44	18

References

Unson, C. G., Macdonald, D., Ray, K., Durrah, T. L., and Merrifield, R. B. (1991) *J. Biol. Chem.*, **266**, pp. 2763-2766.

Unson, C. G., Wu, C.-R., and Merrifield, R. B. (1994) *Biochemistry*, **33**, pp. 6884-6887.

Unson, C. G., Wu, C.-R., Cheung, C. P., and Merrifield, R. B. (1998) *J. Biol. Chem.* **273**, pp. 10308-10312.

Case 27
Regulation of Sugar and Alcohol Metabolism in *Saccharomyces cerevisiae*

Focus concept

The regulation of carbohydrate metabolic pathways in yeast serves as a good model for regulation of the same pathways in multicellular organisms.

Prerequisites

- The major pathways associated with carbohydrate metabolism, including glycolysis, the citric acid cycle, oxidative phosphorylation, pentose phosphate pathway and gluconeogenesis.
- The various fates of pyruvate via alcoholic fermentation and aerobic respiration.

Background

The simple one-celled eucaryotic organism *Saccharomyces cerevisiae* (yeast) has been well studied because extensive research has shown that the metabolic pathways in the yeast are subject to the same types of controls as cells of higher organisms. Yeast are also easy to culture in a laboratory setting. The yeast organism has been employed for literally thousands of years in human history in the production of leavened bread and alcoholic beverages. To the biochemist, the yeast is important because of the Buchner's experiments which disproved the theory of vitalism by demonstrating that metabolic pathways could still occur in cell-free yeast extracts. Their pioneering work allowed scientists to study the metabolic pathways in yeast in great detail. Once pathways had been elucidated, scientists turned their attention to the various mechanisms that served to regulate the pathways. These studies have been carried out with yeast mutants in which a single enzyme associated with a specific pathway is deficient or nonfunctional. Observing the phenotypic character of the mutants allowed scientists to pinpoint the mutated enzyme and increased our understanding of yeast metabolic pathways. In this case, we will focus on carbohydrate metabolism in yeast.

Questions

1. Yeast is used in the production of alcoholic beverages.
   a. Describe how the sugar in the grape (mainly fructose) is converted to ethanol by the yeast cell.
   b. Explain why concentrations of acetate often rise during alcoholic fermentation.

2. It has been observed that cells lacking alcohol dehydrogenase accumulate large amounts of glycerol during anaerobic fermentation. Explain why.

3. Yeast are unusual in that they are able to use ethanol as a gluconeogenic substrate. Ethanol is converted to glucose using the assistance of the glyoxylate pathway. Describe how the ethanol → glucose conversion takes place.

4. A yeast mutant was isolated that was deficient in the enzyme phosphofructokinase. The mutant yeast was able to grow on glycerol as an energy source, but not glucose. Explain why.

5. Yeast is used as a leavening agent in making bread. Explain, in biochemical terms, why the bread dough rises when placed in a warm place.

6. Yeast cells grown on nonfermentable substrates which are then abruptly switched to glucose exhibit *substrate-induced inactivation* of several enzymes. Which enzymes would glucose cause to be inactivated, and why?

7. During anaerobic fermentation, the majority of the available glucose is oxidized via the glycolytic pathway and the rest enters the pentose phosphate pathway to generate NADPH and ribose. This occurs during aerobic respiration as well, except that the percentage of glucose entering the pentose phosphate pathway is much greater in aerobic respiration than during anaerobic fermentation. Explain why.

Reference

Wills, C. (1990) *Critical Reviews in Biochemistry and Molecular Biology*, **25**, pp. 245-280.

Case 28
The Bacterium *Helicobacter pylori* and Peptic Ulcers

Focus concept

The entire genome of *Helicobacter pylori* has recently been sequenced. This will allow biochemists to examine the organism's proteins in detail. In this case, mechanisms employed by *Helicobacter pylori* that allow it to survive in the acidic environment of the stomach will be examined.

Prerequisites

- Protein structure and function.
- Basic metabolic pathways up through amino acid metabolism.

Background

For the past hundred years or so, the cause of peptic ulcers has been attributed to excess acid production by stomach parietal cells. The production of acid was thought to be influenced by environmental factors such as diet and stress. But in 1983, Australian physician Robin Warren described a spiral-shaped bacterium that he had observed on the gastric epithelium of his peptic ulcer patients, and the following year, he and biochemist Barry Marshall put forth the hypothesis that this as-yet unidentified spiral-shaped bacterium was the causative agent of peptic ulcers. Although the scientific community was initially skeptical, further study of the bacterium identified subsequently as *Helicobacter pylori* (or *Campylobacter pylori*) supported this hypothesis. The complete genome sequence of *H. pylori*, reported by Tomb *et al.* in 1997, will allow further detailed studies of the proteins of this bacterium and will provide information that can be used to develop drugs to treat peptic ulcers.

H. pylori has the unusual ability to colonize host cells in the extremely acidic environment of the gastric mucosa, where the pH is typically around 2. *H. pylori* probably employs several different mechanisms to ensure its survival in this environment:

- The bacterium has the ability to establish a positive inside-membrane potential.
- The bacterium may release factors that decrease the secretion of acid by gastric parietal cells.
- The protein urease is essential to the *H. pylori's* ability to survive in the gastric environment.

Questions

1. The proteins found in *H. pylori* have different characteristics than those of bacteria that colonize cells in the usual physiological environment where pH = 7.4. Describe these proteins–what kinds of amino acids would you expect to find in abundance? What would the pI of these proteins be?

2. *H. pylori* synthesizes the enzyme urease, which converts urea to carbon dioxide and ammonia in the aqueous medium of the cell. Write a balanced equation for this reaction. How would the products of this reaction contribute to the survival of *H. pylori* in an acidic environment?

3. A current hypothesis is that *H. pylori* has the ability to establish a *positive inside-membrane potential*. This means that the interior of the bacterium is more positively charged than its exterior. What kinds of mechanisms might *H. pylori* employ in order to accomplish this? List as many as you can think of.

4. The following information concerning the metabolism of *H. pylori* was obtained from the complete genomic sequence:
   - Glucose is the sole source of carbohydrate.
   - Glycolysis, gluconeogenesis, and the pentose phosphate pathway are all active. Lactate dehydrogenase is present.
   - Peptidoglycans (for cell walls) are synthesized from fructose-6-phosphate, phospholipids from glyceraldehyde-3-phosphate and aromatic amino acids from phosphoenolpyruvate.
   - The conversion of pyruvate to acetyl CoA is accomplished by the enzyme pyruvate ferrodoxin oxidoreductase (POR) instead of pyruvate dehydrogenase.
   - The glyoxylate cycle is absent.
   - The tricarboxylic acid cycle is incomplete, ending at isocitrate, which is converted to glutamate in a two-step process.
   - Fatty acid synthetic processes are present.
   - Glutamine synthetase and glutamate dehydrogenase are present.
   - Glutamine is the nitrogen donor in pyrimidine biosynthesis.

   Using the above information, construct a diagram which integrates the major metabolic pathways found in *H. pylori*.

References

Doolittle, R. F. (1997) *Nature* **388**, pp. 515-516.

Tomb, J.-F., *et al.* (1997) *Nature* **388**, 539-547.

Case 29
Pseudovitamin D Deficiency

Focus concept

An apparent Vitamin D deficiency is actually caused by a mutation in an enzyme leading to the vitamin's synthesis.

Prerequisites

- Vitamins and coenzymes.
- The genetic code.

Background

Your patient is a 2-year-old male infant named Justin N. He is suffering from hypotonia, weakness and growth failure, and is unable to walk. His mother has just brought him into the emergency room from the family beach house, where they have been spending the summer, because he has had a seizure. X-rays indicate that the infant is suffering from rickets, which is a result of a nutritional deficiency of Vitamin D. But the infant's mother insists that her son's diet is not Vitamin D-deficient. He drinks three glasses of milk a day, and his diet also includes meat and eggs.

You decide to carry out further analysis and take a sample of the infant's blood. The laboratory results are shown in Table 29.1.

Table 29.1: Laboratory results from a patient with a suspected Vitamin D deficiency.

	Patient	Normal Range
Serum Calcium, mg/dL	5.1	8.7-10.1
Serum Phosphorous, mg/dL	4.2	2.4-4.3
Serum 1α, 25-dihydroxycholecalciferol, pg/mL	13	20-76
Serum 25-hydroxycholecalciferol, ng/mL	48	10-55

A simplified scheme of Vitamin D metabolism is shown in Figure 29.1. The chemical name of active Vitamin D is 1α,25-dihydroxycholecalciferol, and it is synthesized via the pathway shown. Catalysts, both in the form of enzymes and ultraviolet light, are required for Vitamin D synthesis. The two main sources of active Vitamin D are diet and sunlight. Food supplemented with "Vitamin D" usually contains cholecalciferol (Vitamin D_3), or possibly a biologically equivalent analog. In the liver, dietary chole-calciferol is converted to 25-hydroxycholecalciferol. Next, in the kidney, the 25-hydroxycholecalciferol is converted to the active Vitamin D. Sunlight is also responsible for producing Vitamin D_3. The skin contains a precursor, 7-dehydrocholesterol. In the presence of ultraviolet light, which acts as a catalyst, a ring-opening reaction occurs which is followed by the spontaneous conversion of this intermediate to Vitamin D_3. Vitamin D_3 is then converted to active Vitamin D via the pathway just described.

Active Vitamin D is a steroid-like compound that acts in combination with other hormones to increase the concentration of Ca^{2+} via a variety of mechanisms, one of which includes increasing the intestinal absorption of dietary calcium (intestinal absorption of dietary phosphate, a calcium counter-ion, also increases as a result). Calcium ions are required to form hydroxyapatite, $Ca_5(PO_4)_3OH$, the main mineral constituent of bone.

Questions

1. After obtaining the results from the laboratory, you suspect that your patient might have a defective enzyme in the Vitamin D synthetic pathway. Which enzyme do you think is defective, and why?

2. Next, you and your colleagues attempt to isolate the gene coding for the defective enzyme. The gene sequence is shown in Table 29.2. You compare the sequence of the gene from your patient with three other patients you have had with the same symptoms. What is the amino acid change in the enzyme from your patient? What amino acid changes are associated with the enzymes from the other three patients?

Patient	Base Pair Location	Mutation
Justin	319	G → A
Patient A	374	G → A
Patient B	1004	G → C
Patient C	1144	C → T

3. Once you have isolated the mutated gene from your patient, you wish to demonstrate that the gene does in fact code for a nonfunctional protein. You introduce the cloned gene into an expression vector, and these cells produce the protein of interest.

   Design an experiment in which you test the enzymatic activity of your gene product. Assume that you have cultured cells expressing the cloned mutated gene that you can use for this assay. You also have a culture of control cells. Describe the expected results.

4. Explain why you think that the amino acid changes listed in Question 4 would lead to non-functional enzymes.

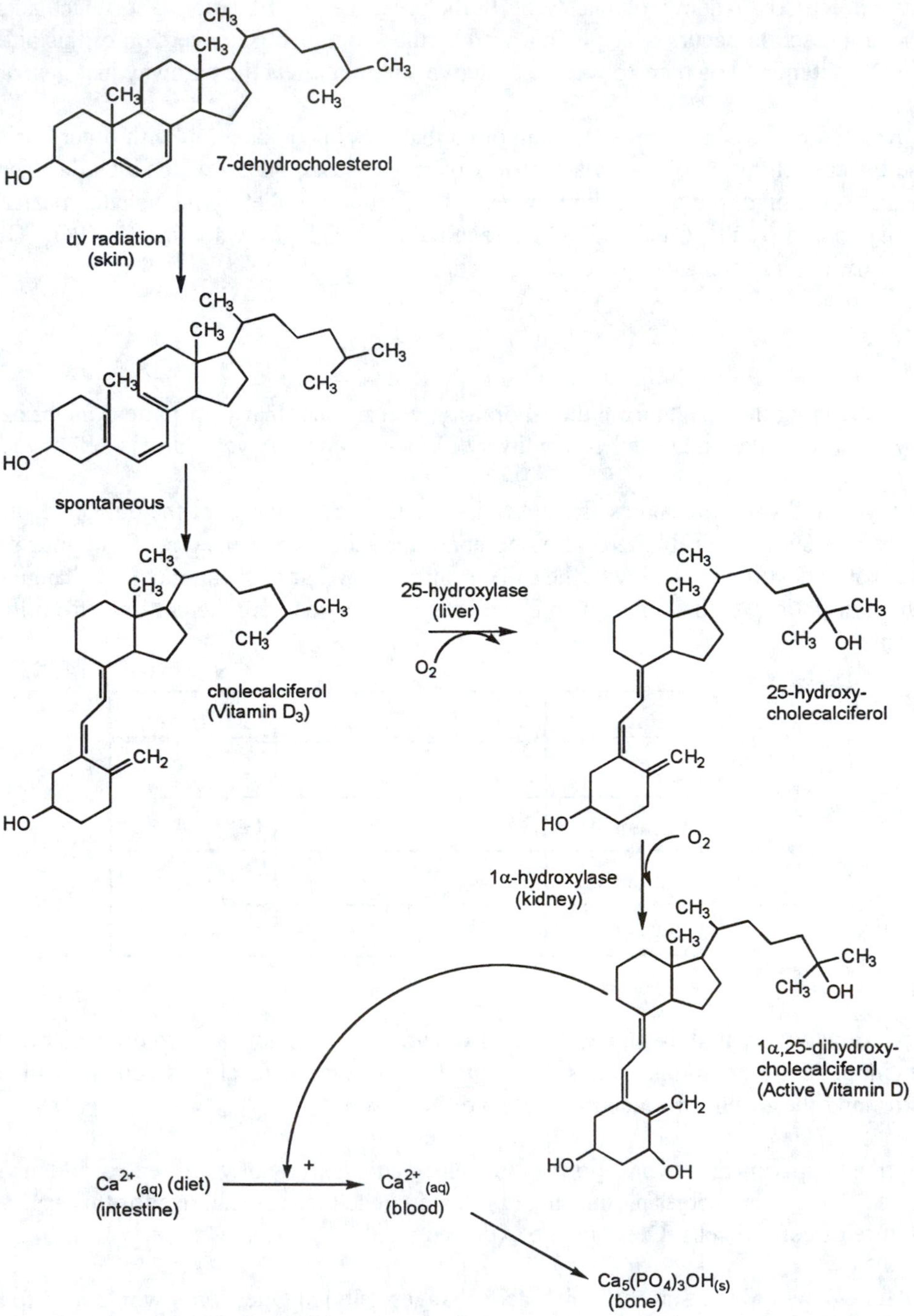

Figure 29.1: Vitamin D metabolism.

92

Table 29.2: The nucleotide sequence of the defective enzyme in pseudovitamin D deficiency.

```
ATGACCCAGA  CCCTCAAGTA  CGCCTCCAGA  GTGTTCCATC  GCGTCCGCTG  GGCGCCCGAG  60
TTGGGCGCCT  CCCTAGGCTA  CCGAGAGTAC  CACTCAGCAC  GCCGGAGCTT  GGCAGACATC  120
CCAGGCCCCT  CTACGCCCAG  CTTTCTGGCC  GAACTTTTCT  GCAAGGGGGG  GCTGTCGAGG  180
CTACACGAGC  TGCAGGTGCA  GGGCGCCGCG  CACTTCGGGC  CGGTGTGGCT  AGCCAGCTTT  240
GGGACAGTGC  GCACCGTGTA  CGTGGCTGCC  CCTGCACTCG  TCGAGGAGCT  GCTGCGACAG  300
GAGGGACCCC  GGCCCGAGCG  CTGCAGCTTC  TCGCCCTGGA  CGGAGCACCG  CCGCTGCCGC  360
CAGCGGGCTT  GCGGACTGCT  CACTGCTGAA  GGCGAAGAAT  GGCAAAGGCT  CCGCAGTCTC  420
CTGGCCCCGC  TCCTCCTCCG  GCCTCAAGCG  GCCGCCCGCT  ACGCCGGAAC  CCTGAACAAC  480
GTAGTCTGCG  ACCTTGTGCG  GCGTCTGAGG  CGCCAGCGGG  GACGTGGCAC  GGGGCCGCCC  540
GCCCTGGTTC  GGGACGTGGC  GGGGGAATTT  TACAAGTTCG  GACTGGAAGG  CATCGCCGCG  600
GTTCTGCTCG  GCTCGCGCTT  GGGCTGCCTG  GAGGCTCAAG  TGCCACCCGA  CACGGAGACC  660
TTCATCCGCG  CTGTGGGCTC  GGTGTTTGTG  TCCACGCTGT  TGACCATGGC  GATGCCCCAC  720
TGGCTGCGCC  ACCTTGTGCC  TGGGCCCTGG  GGCCGCCTCT  GCCGAGACTG  GGACCAGATG  780
TTTGCATTTG  CTCAGAGGCA  CGTGGAGCGG  CGAGAGGCAG  AGGCAGCCAT  GAGGAACGGA  840
GGACAGCCCG  AGAAGGACCT  GGAGTCTGGG  GCGCACCTGA  CCCACTTCCT  GTTCCGGGAA  900
GAGTTGCCTG  CCCAGTCCAT  CCTGGGAAAT  GTGACAGAGT  TGCTATTGGC  GGGAGTGGAC  960
ACGGTGTCCA  ACACGCTCTC  TTGGGCTCTG  TATGAGCTCT  CCCGGCACCC  CGAAGTCCAG  1020
ACAGCACTCC  ACTCAGAGAT  CACAGCTGCC  CTGAGCCCTG  GCTCCAGTGC  CTACCCCTCA  1080
GCCACTGTTC  TGTCCCAGCT  GCCCCTGCTG  AAGGCGGTGG  TCAAGGAAGT  GCTAAGACTG  1140
TACCCTGTGG  TACCTGGAAA  TTCTCGTGTC  CCAGACAAAG  ACATTCATGT  GGGTGACTAT  1200
ATTATCCCCA  AAAATACGCT  GGTCACTCTG  TGTCACTATG  CCACTTCAAG  GGACCCTGCC  1260
CAGTTCCCAG  AGCCAAATTC  TTTTCGTCCA  GCTCGCTGGC  TGGGGGAGGG  TCCCACCCCC  1320
CACCCATTTG  CATCTCTTCC  CTTTGGCTTT  GGCAAGCGCA  GCTGTATGGG  GAGACGCCTG  1380
GCAGAGCTTG  AATTGCAAAT  GGCTTTGGCC  CAGATCCTAA  CACATTTTGA  GGTGCAGCCT  1440
GAGCCAGGTG  CGGCCCCAGT  TAGACCCAAG  ACCCGGACTG  TCCTGGTACC  TGAAAGGAGC  1500
ATCAACCTAC  AGTTTTTGGA  CAGATAG                                          1527
```

Reference

Kitanaka, S., Takeyam, K, Murayama, A., Sato, T., Okumura, K., Nogami, M., Hasegawa, Y., Niimi, H., Yanagisawa, J., Tanaka, T., and Kato, S. (1997) *N. Eng. Jour. Med.*, **338**, pp. 653-661.

Case 30
Phenylketonuria

Focus concept

The characteristics of phenylalanine hydroxylase, the enzyme missing in persons afflicted with the genetic disorder phenylketonuria (PKU), are examined.

Prerequisites

- Amino acid synthesis and degradation pathways.
- Integration of amino acid metabolic pathways with carbohydrate metabolic pathways.

Background

Phenylketonuria is an inherited disease which results from the lack of the enzyme phenylalanine hydroxylase (PAH). The PAH enzyme catalyzes the first step in the degradation of phenylalanine, as shown in Figure 30.1. In the phenylketonuric patient, phenylalanine accumulates which is eventually transaminated to phenylpyruvate, a phenylketone compound. Excess phenylpyruvate accumulates in the blood and urine and has the effect of causing mental retardation if untreated. Screening programs identify PKU babies at birth, and treatment consists of a low phenylalanine diet until maturation of the brain is completed. The structure and biochemical properties of the PAH enzyme have been well-studied.

The gene for PAH has been isolated and has been localized to chromosome 12. The PAH enzyme is a protein 451 amino acids in length with a molecular weight of 51,900 daltons. More than 60 different mutant genes giving rise to nonfunctional PAH proteins have been identified in PKU patients.

Questions

1. Is phenylalanine glucogenic, ketogenic, or both? Explain.

2. In order to learn more about the PAH enzyme, it was necessary to purify it. PAH has been isolated from both rats and humans. In the rat, three isozymes of PAH have been identified in the liver. Their molecular weights are identical, but their charges are different, as demonstrated by isoelectric focusing. The pI values are 5.2, 5.3 and 5.6. DEAE-cellulose (anion exchange) chromatography was one of the steps in the purification procedure of the enzymes. Predict the order of elution of these isozymes from the DEAE-cellulose column. What pH buffer would you choose in running the column?

3. Once the enzyme was purified, the investigators set out to determine its properties. They wanted to see if phenylalanine, in addition to serving as a substrate for the enzyme, had an additional role in the regulation of the enzyme. Polyacrylamide gels (under denaturing and non-denaturing conditions) were carried out with the rat liver PAH. The results are shown in Figure 30.2. How do you interpret these data?

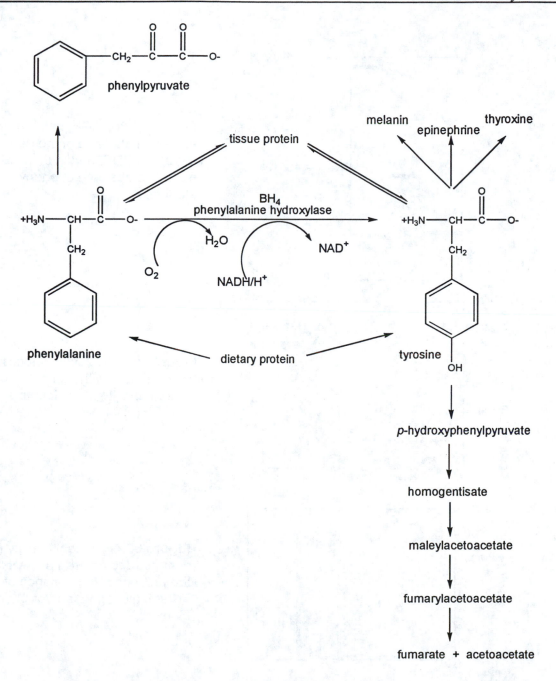

Figure 30.1. Phenylalanine and tyrosine metabolism. BH$_4$ is tetrahydrobiopterin, an essential cofactor for phenylalanine hydroxylase.

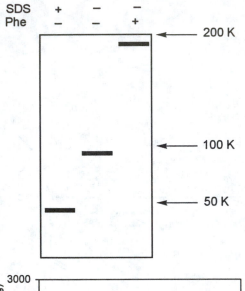

SDS + – –
Phe – – +

← 200 K

← 100 K

← 50 K

Figure 30.2: PAGE, in the presence and absence of the denaturing agent SDS, and in the presence and absence of phenylalanine.

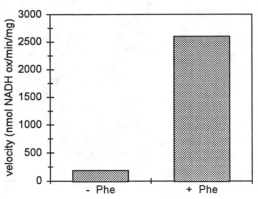

Figure 30.3: Reaction kinetics of PAH with and without pre-incubation with phenylalanine.

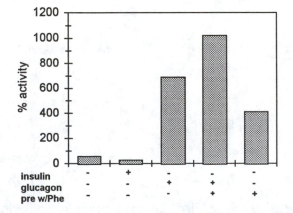

Figure 30.4: Percent activity (as compared to control) of PAH in the presence and absence of insulin, glucagon, and/or phenylalanine.

4. Next, kinetic studies were carried out with the enzymes. A plot of velocity vs. phenylalanine concentration yields a sigmoidally shaped curve. What does this tell you about the enzyme?

5. Kinetic data in which PAH activity is compared with and without preincubation with phenylalanine is shown in Figure 30.3. Give a structural basis for the interpretation of these data.

6. The effect of the hormones glucagon and insulin on PAH activity were investigated. The results are shown in Figure 30.4. In addition, the amount of radioactively-labeled phosphate incorporated into PAH with glucagon treatment was found to be nearly seven-fold greater than in controls.
   a. How would you interpret these data?
   b. Draw a diagram demonstrating the mechanism for hormonal activation of PAH.
   c. Which hormone activates PAH, and why?

7. Tyrosine is not an essential amino acid in normal persons, but it is essential in persons with PKU. Explain why.

8. Patients with the disease PKU tend to have blue eyes, fair hair, and very light skin. Explain why.

9. More than 60 different mutant PAH genes have been identified in PKU patients. State, in general, what effect the following changes in the DNA would have on the resulting protein. Be specific.
   a. Nonsense mutations
   b. Splicing mutations
   c. Single base changes

10. The cause for the mental retardation associated with untreated PKU is not completely understood, but it is believed to arise from the high concentrations of phenylpyruvate, which is a product of a transamination reaction with phenylalanine and α-ketoglutarate. The phenylpyruvate is believed to be toxic to the developing brain. Write the balanced equation for the transamination of phenylalanine to phenylpyruvate, and include the structures of the reactants and products. Identify any cofactors needed to accomplish the reaction.

11. The mental retardation associated with phenylalanine can be avoided if the neonate is immediately placed on a low phenylalanine diet for the early years, and perhaps for life.
   a. Why is a PKU patient placed on a *low phenylalanine* diet instead of a *phenylalanine free* diet?
   b. The artificial sweetener Nutrasweet ® contains the compound aspartame, which consists of a methylated Asp-Phe dipeptide. (The C-terminal carboxyl group is methylated.) Draw the structure of aspartame. If you were a physician, what advice would you give to a PKU patient regarding this product? If you were a manufacturer of aspartame, what would your responsibility to your customers be?

References

Agranoff, B. W., and Aprison, M. H. (1977) *Advances in Neurochemistry*, Plenum Press, NY, pp. 1-132.

Kaufman, S. (1997) *Tetrahydrobiopterin: Basic Biochemistry and Role in Human Disease*, The Johns Hopkins University Press, Baltimore, MD, pp. 31-139 and 262-322.

DiLella, A. G., Marvit, J., and Woo, S. L. C. (1987) The Molecular Genetics of Phenylketonuria in *Amino Acids in Health and Disease: New Perspectives*, Alan R. Liss, Inc., pp. 553-564.

Case 31
Hyperactive DNAse I Variants:
A Treatment for Cystic Fibrosis

Focus concept

Understanding the mechanism of action of an enzyme can lead to the construction of hyperactive variant enzymes with a greater catalytic efficiency than the wild type enzyme.

Prerequisites

- Enzyme kinetics and inhibition.
- DNA structure.
- The hyperchromic effect.
- The properties of supercoiled DNA.

Background

The enzyme deoxyribonuclease I (DNAse I) is an endonuclease that hydrolyzes the phosphodiester bonds of the double-stranded DNA backbone to yield small oligonucleotide fragments. DNAse I is used therapeutically to treat patients with cystic fibrosis (CF). The DNAse I enzyme is inhaled into the lungs where it then acts upon the DNA contained in the viscous sputum secreted by the lungs in these patients. Hydrolysis of high molecular weight DNA to low molecular weight DNA in the sputum decreases its viscosity and improves lung function. Animal studies also have shown that DNAse I is effective in treating the autoimmune disease systemic lupus erythematosus (SLE). In this disease, the DNA secreted into the serum provokes an immune response. DNAse I prevents the immune response by degrading the DNA to smaller fragments that are not recognized by the immune system.

Genentech, Inc., the company that produces the recombinant DNAse I, was interested in improving the efficiency of DNAse I so that less drug would be needed to achieve the same results. Scientists in the protein engineering lab constructed hyperactive variants at DNAse I which actually worked better than the wild-type enzyme. DNAse I acts by processively *nicking* the phosphodiester backbone, so the scientists reasoned that a variant that could create more nicks in a shorter period of time would act more efficiently than the wild-type enzyme. In this case, we will examine the engineered hyperactive variants and use the results to make some conclusions about the mechanism of DNAse I.

Questions

1. The DNAse I variants engineered by the Genentech scientists are listed in Table 31.1. (A note on nomenclature: Q9R mean that the glutamine at position 9 in the wild type DNAse I enzyme has been changed to an arginine.)
   a. What structural feature do all of the DNAse I variants have in common? Explain the meaning of the abbreviation in the table.

b. Why do you suppose that the protein engineers thought that these changes would improve the catalytic efficiency of DNAse I?

Table 31.1: DNAse I variants.

Variant	Abbreviation
Q9R	+1
E13R	
T14K	
H44K	
N74K	
T205K	
E13R/N74K	+2
Q9R/E13R/N74K	+3
E13R/N74K/T205	
Q9R/E13R/N74K/T205K	+4
E13R/H44K/N74K/T205K	
T14K/H44K/N74K/T205K	
E13R/T14K/N74K/T205K	
Q9R/E13R/H44K/N74K/T205K	+5
Q9R/E13R/T14K/H44R/N74K/T205K	+6

2. The enzymatic activity of the DNAse I variants was tested using a DNA hyperchromicity assay. The absorbance of a solution of intact DNA was measured at 260 nm. Then the enzyme was added, and the solution was monitored for an increase in absorbance. (The increase in absorbance at 260 nm is referred to as the hyperchromic effect.) Why was a hyperchromicity assay effective in assessing the activity of the DNAse I variants?

3. The DNA hyperchromicity assay was used to measure the K_M and v_{max} values for each variant. The results are shown in Table 31.2. Explain the significance of the K_M and v_{max} values. What effect has the amino acid change(s) had on the activity of the enzyme variants as compared to the wild type?

Table 31.1: DNAse I variants. (Based on Pan and Lazarus, 1998.)

Variant	K_M, μg/mL DNA	v_{max}, A_{260} units/min/mg DNAse I
Wild type	1.0	1.0
Q9R	1.1	2.8
E13R	0.23	1.5
T14K	0.43	1.1
H44K	0.43	1.1
N74K	0.77	3.6
T205K	0.42	2.1
E13R/N74K	0.20	5.3
Q9R/E13R/N74K	0.20	7.0
E13R/N74K/T205	0.18	7.7
Q9R/E13R/N74K/T205K	0.09	2.8
E13R/H44K/N74K/T205K	0.17	6.4
T14K/H44K/N74K/T205K	0.18	7.7
E13R/T14K/N74K/T205K	0.37	3.5
Q9R/E13R/H44K/N74K/T205K	0.11	2.4
Q9R/E13R/T14K/H44R/N74K/T205K	0.20	2.5

4. Next, the protein engineers wished to characterize the DNAse I variants in terms of their ability to *cut* or *nick* DNA. A *cut* refers to the hydrolysis of phosphodiester bonds on both strands, whereas a *nick* is the hydrolysis of just one strand. This was assessed by using the circular plasmid pBR322. The plasmid is the most stable in the supercoiled form. If the phosphodiester backbone is nicked on one strand, the plasmid forms a relaxed circle, but if the backbone is cut on both strands, the circle linearizes, as shown in Figure 31.1. Supercoiled, relaxed circular and linear DNA can be detected by differential migration through agarose gels. In a series of experiments, pBR322 substrate was incubated with DNAse I for 45 minutes, then the products were analyzed by agarose gel electrophoresis. The results are shown in Figure 31.2. Describe the results for each lane. Compare the selected variants with the wild type DNAse I with regard to their ability to cut or nick the DNA.

101

Figure 31.1: Conversion of supercoiled DNA to relaxed and linear DNA via phosphodiester bond hydrolysis.

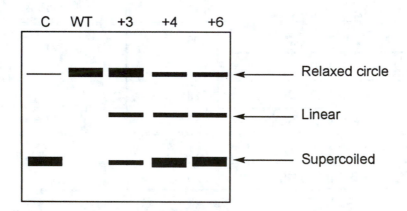

Figure 31.2: Supercoiled plasmid DNA digestion by DNAse I variants and analysis by agarose gel electrophoresis. C is the control, WT is wild type DNAse I, +3 is 13R74K205K, +4 is 13R14K74K205K and +6 is 9R13R14K44R74K205K. (Based on Pan and Lazarus, 1998.)

5. The investigators next tested the DNAse I variants' enzymatic ability with high and low molecular weight DNA, in high and low concentrations. High molecular weight DNA is present in the lung secretions of CF patients in fairly high concentrations, but the DNA present in the serum of mice with SLE is present in one-tenth the concentration. The data are shown in Table 31.3. What is your interpretation of these data?

Table 31.3: Dependence of DNA nicking activity by DNAse I variants on DNA length and concentration. (Based on Pan and Lazarus, 1998.)

	Relative nicking activity (as compared to wild type)			
	Low DNA conc		High DNA conc	
	Low mwt	*High mwt*	*Low mwt*	*High mwt*
Wild type	1	1	1	1
N74K	26		10.4	
E13R/N74K	211	31	24.3	13
E13R/N74K/T205K	7		1.3	
E13R/T14K/N74K/T205 K	7		0.7	

6. Use the experimental evidence presented here to compare the mechanism of the wild type DNAse I with the variant DNAse I enzymes.

Reference

Pan, C. Q., and Lazarus, R. A. (1998) *J. Biol. Chem.* **273**, pp. 11701-11708.

Answers to
Selected
Cases

Case 3
Carbonic Anhydrase II Deficiency

1.

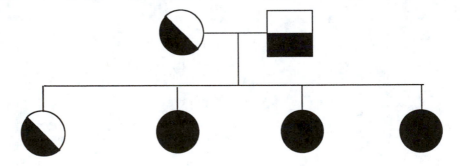

2. The parents and the unaffected sister are all heterozygotes. The parents have one normal carbonic anhydrase II gene and one defective gene. The unaffected sister has inherited one normal gene from one parent and one defective gene from the other parent. The three affected sisters have inherited the defective gene from each parent.

3. The histidine #107 is not one of the histidines involved in binding to the zinc ion in the catalytic site of the enzyme. However, the imidazole side chain of the histidine must be important in maintaining the proper three-dimensional conformation of the enzyme. The tyrosine amino acid side chain consists of a phenol side chain and has different hydrogen-bond forming capabilities than the imidazole side chain. The imidazole side chain can also be protonated, which would give it a positive charge which would be capable of forming ion pairs with negatively charged amino acid side chains. Tyrosine does not have this capability.

4. Hydrogen ions are produced from carbonic acid, which was produced from water and carbon dioxide via the carbonic anhydrase II reaction. The hydrogen ions exit the cell and enter the bone-resorbing compartment via the H^+/Na^+ exchanger. Hydrogen ions leave the osteoclast to acidify the bone-resorbing compartment and sodium ions enter. Next, sodium ions leave the cell via the Na^+/K^+ exchanger while potassium ions come into the cell. Bicarbonate, the second product of the dissociation of carbonic acid, leaves the cell via the HCO_3^-/Cl^- exchanger in which bicarbonate leaves and chloride enters the cell.

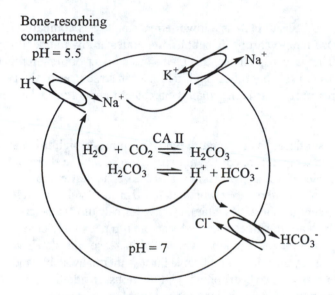

Case 5
Characterization of Subtilisin from the Antarctic Psychrophile *Bacillus* TA41

1. a. The investigators would have used SDS-PAGE to determine the protein's molecular weight. They would have used isoelectric focusing to determine the protein's pI.

    b. The pI of subtilisin S41 is much lower than the pI of mesophilic subtilisin. This indicates that the subtilisin S41 has a greater number of acidic residues (such as Asp and Glu) than the mesophilic subtilisins. In fact, subtilisin S41 has 2-3 times more Asp residues than common mesophilic subtilisins.

2. The psychrophilic subtilisin S41 has a low thermal stability as compared to the mesophilic subtilisin Carlsberg. Because the psychrophilic subtilisin operates at low temperatures, a high thermal stability is not required. The psychrophilic subtilisin also has a lower optimal temperature of activity than the mesophilic subtilisin, probably reflecting differences in structures of the two enzymes. (Interestingly, the optimal temperature is far higher than the ambient temperature where both these organisms are found.)

3. The thermophilic subtilisin has a structure which promotes a high thermal stability. This is required because of the high temperatures of the hot springs in which the thermophilic bacteria reside. The high stability is achieved via a small compact structure, a great number of ionic interactions and aromatic (hydrophobic) interactions which increase protein stability as described in the background. The thermophilic subtilisin also has a high affinity for calcium, as described by its low K_d value. Calcium ions, which are positively charged, can bind to negatively charged amino acid side chains in the protein and form ionic interactions, which also increases the stability of the protein.

    In contrast, the psychrophilic subtilisin is far less thermally stable and thus has fewer ionic interactions and no aromatic (hydrophobic) interactions. In addition, the psychrophilic subtilisin has a low affinity for calcium, so calcium ions are not involved in ionic interactions. In addition, the psychrophilic subtilisin is a much larger protein, which allows it to be more flexible and less compact. The increased number of Asp residues, which are hydrophilic, prefer to reside on the surface of the protein. This gives the psychrophilic subtilisin a greater surface area, which makes the protein structure more open and less compact. This open flexible structure will be able to accommodate the substrate in the enzyme's active site and will help solve the problem of decreased reaction rates associated with low temperatures.

Case 7
A Storage Protein from Seeds of *B. nigra* is a Serine Protease Inhibitor

1. a. Gel filtration chromatography separates proteins on the basis of molecular size, which is generally proportional to molecular weight.

   b. The BN protein must be larger than the other proteins in the *B. nigra* seeds, since it elutes from the column first.

2. In the dialysis procedure, the protein is placed inside dialysis tubing. The pores in the dialysis tubing allow proteins with molecular weights smaller than 6000-8000 daltons to pass through into the dialysis buffer, while proteins larger than this, such as the BN, are retained. Thus, the BN solution in the dialysis bag is rid of small molecular weight contaminant proteins.

3. a.

$$CH_3 - \underset{\underset{O}{\|}}{C} - NH - \underset{\underset{\underset{\underset{OH}{|}}{CH_2}}{|}}{CH} - COO^-$$

   b. The Edman reagent requires a primary amine group to react with. If an acetyl group is blocking the α-amino group on the N-terminal of the protein, the initial reaction with the Edman reagent cannot take place and the fragment cannot be sequenced using this method.

4. a. If only one proteolytic cleavage was carried out, the sequences of the fragments could be determined but you would not know how to place the fragments in the proper order. If two proteolytic cleavages are carried out using two different enzymes, *overlap peptides* will result and the fragments can be placed in the proper order.

   b. Trypsin cleaves after Lys and Arg residues. The fragments resulting from digestion of the heavy and light chains are shown below, identified by amino acid residue number.

Light chain	Heavy chain
6	1-7
7-9	8-18
10-11	19-31
12	32-35
13-21	36-38
22-29	39-42
30-39	43-57
	58-61
	62-71
	72-77
	78-85
	86-91

c. Chymotrypsin would be a good choice to use for the second protease. Chymotrypsin cleaves after the aromatic amino acids Phe, Tyr, and Trp. The following fragments would result:

Light chain	Heavy chain
6-14	1-63
15-26	64-83
27-39	84-91

5. The protein is unusually stable. It appears to retain about 65% of its secondary structure, even at temperatures as high as 80°C. Even at 100°C, the boiling point of water, 50% of the secondary structure still remains.

6. If BN is a competitive inhibitor, it must bind to the active site of the enzyme and prevent substrate from binding. Competitive inhibitors resemble their substrates (since both fit into the active site), so it's not surprising that the inhibitor would be a protein, since the substrates are also proteins. The serine proteases have an identical catalytic triad, but different specificity "pockets", which is why they can carry out the same reaction (cleavage of a peptide bond) but do so after different amino acids. If BN inhibits all three enzymes, it probably binds near the catalytic triad, since this is the structural feature that all three of the enzymes have in common.

7. Since ions cannot quench the fluorescence, the tryptophan must reside in a nonpolar environment where hydrated, charged ions cannot penetrate. The total lack of quenching by the positively charged cesium ion might mean that there are positive amino acid side chains in the vicinity of the tryptophan, which would repel the cesium ion. The small, nonpolar acrylamide was able to penetrate the protein molecule to contact tryptophan, but its quenching efficiency was also low, perhaps because there was not sufficient space in the protein to accommodate a molecule even as small as acrylamide.

1.

$$+H_3N-\overset{\overset{\displaystyle O}{\parallel}}{\underset{\displaystyle CH}{|}}-C-O^- + NADPH + H^+ \xrightarrow{\text{NOS}} NO + {}^+H_3N-\overset{\overset{\displaystyle O}{\parallel}}{\underset{\displaystyle CH}{|}}-C-O^- + NADP^+$$

$$+ 2\,O_2 \qquad\qquad\qquad + 2\,H_2O$$

For arginine: the side chain is CH_2—CH_2—CH_2—NH—$C=NH_2^+$—NH_2

For citrulline: the side chain is CH_2—CH_2—CH_2—NH—$C=O$—NH_2

arginine   citrulline

2. In the R form (oxy), the Cysβ93 is exposed to the solvent containing the NEM reagent and the reaction can proceed. In the T form (deoxy), the Cysβ93 must be buried and therefore unaccessible.

3. The NO reacts with oxy-Hb faster than with deoxyHb, according to the results presented here. Since NO reacts with the Cysβ93 residue, the reaction rate is faster with the oxy-Hb because the Cysβ93 residue is more exposed in the R conformation.

4. The oxygenated arterial blood contains a higher percentage of hemoglobin in the R form and has a high concentration of S-nitrosohemoglobin. In the venous blood, which is deoxygenated and contains a higher percentage of hemoglobin in the T form, the NO is released from the hemoglobin into the circulation. The presence of the SNO-Hb can be attributed to the concentration of NO, since inhibition of NO synthase (NOS) decreases the amount of SNO-Hb found in arterial blood.

5. If purified hemoglobin is injected into the circulation of an experimental animal, its blood pressure increases. This could be the result of scavenging of NO by the exogenous Hb. Once deprived of NO, endothelial cells would not be able to relax and blood pressure would increase as a consequence. Since the SNO-Hb already has NO bound, it would not be able to scavenge endogenous NO.

6. a. In hypoxic tissue, the O_2 partial pressure will be low. This will favor the T form, which has a low affinity for oxygen. Thus, in making the transition from the T to the R form, O_2 is delivered to the hypoxic tissue. However, SNO-Hb prefers the R form, so if Hb is to make the transition to the T form, it must release its NO. Once released, the NO will diffuse into muscle cells and promote relaxation. Thus, in making the transition from the T to the R form, O_2 is delivered to the cells, and the blood vessel relaxes, which improves blood flow and facilitates delivery of the oxygen.

    b. If O_2 were plentiful, hemoglobin would be in the R form with O_2 bound. In the R form, the reaction of hemoglobin to SNO-Hb is favored. This has the effect of scavenging any endogenous NO. Deprived of NO, muscle cells lining the blood vessels constrict, increasing blood pressure.

Case 12
Production of Methanol in Ripening Fruit

1. There was an increase in PME activity as the fruit ripened for the wild-type tomato. For the PME-deficient tomato, there was very little PME activity. The activity was correlated with total protein, as shown by the Western blot–low activity meant that not much protein was being synthesized whereas higher activities were the result of increased protein synthesis. (The exception to this correlation occurred at the RR stage, where protein content remained the same as in the TU stage but PME activity decreased.) Therefore, PME is regulated at the level of translation rather than being inhibited or stimulated by allosteric effectors.

2. a. There is an overall general correlation of PME activity and methanol production. In the wild-type tomato, as PME increases, methanol concentration increases as well. However, there is a short lag time between PME activity and methanol production between the IMG and MG stages where PME activity increased, but methanol production remained the same. There is also an increase in methanol between the TU and RR ripening stages, but the PME activity actually decreases between these two stages. For the transgenic tomato, the methanol content and PME activity both remain low.
   b. These results indicate that the PME enzyme is responsible for methanol production in ripening fruit.

3. Ethanol production in the wild-type tomato is low in all stages of the ripening process, and decreases to even lower levels as the ripening progresses. But in the transgenic tomato, ethanol concentration increases dramatically between the TU and RR stages. The production of ethanol and methanol appear to be inversely correlated. In the wild-type tomato, production of ethanol decreases as methanol increases. In the transgenic tomato, ethanol production increases in the absence of methanol production.

4.

$$\underset{\text{Acetaldehyde}}{\overset{\displaystyle O}{\underset{CH_3}{\overset{\|}{C}}-H}} + NADH + H^+ \xrightarrow{\ ADH\ } \underset{\text{Ethanol}}{\overset{OH}{\underset{CH_3}{\overset{|}{\underset{|}{CH_2}}}}} + NAD^+$$

5. It is likely that methanol serves as an inhibitor for ADH, since ethanol levels are decreased in the wild-type tomato when methanol levels are at their highest. In the transgenic tomato, there is no methanol production in the absence of PME, so there is nothing to inhibit the ADH enzyme and consequently ethanol levels increased.

Case 15
Site-Directed Mutagenesis of Creatine Kinase

1. The Cys 278 is highly exposed and unusually reactive as compared to other cysteines in the creatine kinase enzyme. The Cys 278, because of its high reactivity, is probably one of the catalytic residues in the enzyme. The other cysteine residues are probably not as reactive because they are not directly involved in catalysis and/or because they are shielded in some way which prevents them from reacting with the NEM reagent.

2. a. Since Gly, Ser, Ala, Asn and Asp all lack the sulfhydryl group, the sulfhydryl group (or some chemical property that the sulfhydryl group possesses) must be essential to the activity of the enzyme.

   b. The difference between C278D and C278N is that the mutants contain Asp and Asn residues, respectively. The main difference between these two amino acid side chains is the negative charge on the Asp, while the Asn is neutral. Since the C287D mutant has a greater activity, the negative charge must be essential to the enzyme's catalytic activity.

   c. Chloride and bromide ions are negatively charged. It's possible that a negative charge at the 278 site is important to the catalytic mechanism of the creatine kinase enzyme. The cysteine could provide this negative charge through deprotonation of the -SH group to yield -S$^-$. Without the Cys residue, a negative charge cannot form, so it's possible that the chloride and bromide ions provide this needed negative charge, which would account for the enhancement seen in the enzymatic activity.

   d. The C278D mutant contains a negatively charged Asp residue. The negatively charged bromide and chloride ions would repel the side chain on the Asp residue and cause the enzyme activity to decrease.

3. The iodoacetate-modified enzyme preserves the negative charge whereas in the iodoacetamide-modified enzyme, the negative charge is abolished. Since the current study indicates that the negative charge is essential, the iodoacetate-modified enzyme retains some enzymatic activity because of the presence of the negative charge on the iodoacetate-modified cysteine side chain.

4. a. The wild-type enzyme has the highest maximal velocity of the three enzymes presented here and it also has the highest affinity for both creatine and ATP substrates. The wild-type enzyme has a higher affinity for ATP as a substrate than for creatine. For both substrates, the K_M values are lower than the K_d values. This indicates that, for both creatine and ATP, the binding of the second substrate is facilitated by the binding of the first substrate. Thus, the binding of substrates is synergistic–the binding of one substrate allows the second substrate to bind more easily.

   b. For both the C278G mutant and the C278S mutant, the K_M and K_d values are higher for both creatine and ATP than the wild-type, indicating that the affinity for both substrates is less for the mutant than for the wild-type enzyme. The mutant binds ATP with greater affinity than creatine. However, when the K_d and K_M values are compared, the opposite pattern is observed as compared to the wild-type: For both substrates, the K_M values are greater than the K_d values. This indicates that the binding of the second substrate is not facilitated; in fact it's the opposite.

Thus when Cys 278 is mutated, the synergistic binding of the two substrates is lost, indicating that it's likely that the Cys278 is responsible for the synergistic effect. Perhaps the binding of the first substrate causes a conformational change in the mutant protein which makes the binding of the second substrate unfavorable.

c. The V_{max} values are decreased in the mutant enzymes, and the synergistic effect of substrate binding is lost when the Cys 278 is mutated. Thus, the sulfhydryl group on the side chain of the cysteine may play a role in the synergistic effect of substrate binding, and this binding is essential to the optimum functioning of the enzyme.

Case 17
A Possible Mechanism for Blindness Associated with Diabetes
Sodium-Dependent Glucose Uptake by Retinal Cells

1. Glucose uptake by pericytes nearly doubles when sodium is present. In contrast, in endothelial cells, the uptake of glucose is constant, regardless of whether sodium is present or absent.

2. Glucose uptake increases as sodium concentration increases in pericytes. In endothelial cells, glucose uptake is constant regardless of sodium ion concentration. The shape of the curve for the pericytes indicates that a protein transporter is involved–glucose uptake increases linearly as sodium ion concentration increases, then reaches a plateau at high sodium ion concentrations, indicating that the transporter is saturated and is operating at its maximal capacity.

3.

	K_M	v_{max}
With sodium ions	$K_M = -1/-0.25 \text{ mM}^{-1}$ $K_M = 4 \text{ mM}$	$v_{max} = 1/0.028 \text{ mg/ hr/nmol}$ $v_{max} = 36 \text{ nmol/hr/mg}$
Without sodium ions	$K_M = -1/-0.25 \text{ mM}^{-1}$ $K_M = 4 \text{ mM}$	$v_{max} = 1/0.063 \text{ mg/ hr/nmol}$ $v_{max} = 16 \text{ nmol/hr/mg}$

The K_M values for glucose transport in the presence and absence of sodium ions are the same, indicating that the transporter has equal affinity for glucose in the presence and absence of sodium ions. However, the maximal velocity is much greater in the presence of sodium ions than in the absence of sodium ions. Because the K_M values are the same, sodium ions probably don't increase the transporter's affinity for glucose. Instead, it is more likely that the sodium ions cause some kind of conformational change which allows the transporter to transport glucose more effectively.

The investigators have convincingly demonstrated that an SGLT exists in pericytes. Glucose transport increases as sodium ion concentration increases. The transport curve is hyperbolic, indicating that a protein transporter which has specific binding sites for glucose exists in pericytes. There is probably not a SGLT in endothelial cells, since glucose transport is independent of sodium ion concentration.

4. a. Galactose and 2-deoxyglucose inhibited glucose transport only slightly in the presence and absence of NaCl.
   b. Phlorizin inhibits glucose transport in the presence of sodium ions, but does not inhibit glucose transport in the absence of sodium ions.

5. When glucose concentration increases, cell density increases, which is due to increased collagen synthesis in pericytes. Thus, increased glucose transport leads to increased collagen synthesis.

6. a. In the absence of phlorizin, glucose transport increases with increasing concentration of glucose. But in the presence of phlorizin, glucose transport decreases.

    b. When glucose concentration increases, the cell density increases in the absence of phlorizin. But when phlorizin is present, glucose transport decreases, as shown in Figure 17.6 (left). In the presence of phlorizin, both Type IV and Type VI collagen synthesis decreases also. Thus, a link has been established between glucose transport and collagen synthesis: When glucose concentration increases, transport into the cell increases, which stimulates collagen production.

7. In the untreated diabetic, plasma glucose concentrations will be high because insulin is not present to stimulate the uptake of glucose into the cells. Pericytes, however, do not rely on insulin for glucose uptake; instead glucose uptake is stimulated by sodium ions. The experimental data show that when glucose concentration is increased, glucose uptake via the SGLT transporter also increases. The results also show that increased glucose uptake leads to increased collagen synthesis, which damages the pericytes and leads to retinopathy.

    In order to prevent this from occurring, diabetics should carefully control their plasma glucose concentration through exogenous insulin and diet. In addition, drugs that act in the same manner as phlorizin might be helpful, since the results show that in the presence of phlorizin, glucose uptake by the SGLT decreases, even in the presence of high glucose concentration.

Case 20
NAD⁺-dependent Glyceraldehyde-3-phosphate Dehydrogenase from *T. tenax*

1. Hexokinase, phosphofructokinase-1, and pyruvate kinase.

2.

Glyceraldehyde-3-phosphate   1,3-bisphosphoglycerate

3. In the GAPDH reaction, NAD⁺ is reduced to NADH and H⁺ and glyceraldehyde-3-phosphate is phosphorylated to 1,3-bisphosphoglycerate. Thus this reaction couples oxidation and phosphorylation. In the next step, ATP will be generated when 1,3-bisphosphoglycerate donates one of its phosphates to ADP. Subsequent reactions (production of lactate, ethanol, or oxidative phosphorylation) must re-oxidize the NADH to NAD⁺ so that the glycolytic pathway can continue.

4. The product, 1,3-bisphosphoglycerate is not a substrate for the GAPDH enzyme. This indicates that the reaction only takes place in the forward direction and not in the reverse direction.

5. a. NADP⁺, NADH and ATP are inhibitors of GAPDH because K_M values are greater than the control, indicating that GAPDH has a low affinity for substrate in the presence of these inhibitors. Glucose-1-phosphate, AMP and ADP are activators of GAPDH because the K_M values are less than the control, indicating a greater affinity for substrate in the presence of these activators.
   b. NADH and ATP are inhibitors of GAPDH because NADH is a product of the reaction and ATP is a product of the glycolytic pathway as a whole. Increased concentrations of NADH and ATP indicate a high energy charge of the cell. Glycolysis is inhibited if NADH and ATP are not needed. In the same manner, AMP and ADP are activators of GAPDH because AMP and ADP indicate that the energy charge of the cell is low and ATP is needed and thus glycolysis is stimulated. Glucose-1-phosphate is an activator because glucose-1-phosphate is the product of glycogenolysis. Glucose-1-phosphate is converted to glucose-6-phosphate, and then the glucose-6-phosphate enters glycolysis. Thus, the glucose-1-phosphate acts via *feed-forward stimulation* and stimulates the activity of a "downstream" enzyme so that additional substrate can be accommodated in the glycolytic pathway. The inhibition by NADP⁺ might be explained by noting that the other GAPDH isozyme is dependent upon NADP⁺. Inhibition of the NAD⁺-dependent isozyme may ensure that both enzymes are not simultaneously active.

6. The NAD⁺-dependent GAPDH is composed of a tetramer and it is quite possible that there are multiple binding sites for its substrates. Figure 20.1 shows that the NADP⁺ inhibition is represented by a sigmoidally shaped curve, indicating cooperativity of binding of the NADP⁺ inhibitor.

Case 22
Carrier-mediated Uptake of Lactate in Rat Hepatocytes

1. The Cori cycle is active during periods of exercise or starvation. During exercise, glucose in the muscle enters glycolysis and is converted to pyruvate, then lactate during anaerobic fermentation. The lactate then diffuses out of the muscle cell and enters the

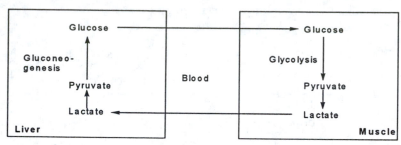

bloodstream where it is taken up by the liver. Lactate then enters gluconeogenesis to form glucose, which leaves the liver and can return to the muscle or other body tissues in need of glucose. The Cori cycle provides a means by which blood glucose concentrations can be maintained in the absence of exogenous glucose.

2. Lactate is negatively charged (it is unprotonated at physiological pH) and would have difficulty passing through the plasma membrane because of its charge and polarity. For these reasons, the transport of lactate is likely to be carrier-mediated.

3. The transport of lactate is mediated by a protein, since the transport curves are hyperbolic. The transported lactate binds to binding sites on the carrier protein, so the rate of transport is proportional to lactate concentration at low concentrations of lactate. But at high concentrations of lactate, the rate of transport reaches a plateau because all of the binding sites are occupied. The transporter has a preference for L-lactate, as L-lactate is transported more rapidly than D-lactate. This confirms the presence of binding sites on the carrier protein for L-lactate. D-lactate is likely to have difficulty binding to sites specific for L-lactate and thus is transported less efficiently.

4. Transport is the most efficient when the external pH is low. These data can be interpreted in two different ways–one is that the lactate is preferentially transported in the protonated form. The other possibility is that the lactate is transported in unprotonated form in symport with a hydrogen ion.

5. In the presence of high concentrations of sulfate ions, lactate was still transported at a rate of 89% of normal. This indicates that lactate and sulfate are transported by different transporters. If lactate and sulfate shared the same transporter, then sulfate would inhibit lactate transport because the sulfate ions would occupy the vast majority of the binding sites since sulfate is present at a high concentration. Since lactate transport is not appreciably inhibited in the presence of sulfate ions, it is likely that lactate and sulfate are transported by distinct transport proteins.

Case 24
Uncoupling Proteins in Plants

1. Uncoupling agents work by dissipating the proton gradient between the mitochondrial matrix and the inner membrane space of the mitochondrion. The proton gradient serves as the energy reservoir which couples flow of electrons through the electron transport chain to ATP synthesis by the ATP synthase enzyme. If these two processes are uncoupled due to the loss of the proton gradient, electron transport still occurs but ATP synthesis does not occur. The energy released during electron transport is released in the form of heat if the energy cannot be used to synthesize ATP.

2. In the root system of the skunk cabbage, starch is broken down enzymatically to yield glucose. Glucose is then broken down via aerobic respiration via glycolysis, the citric acid cycle and the electron transport chain. The oxidation of glucose provides the $NADH/H^+$ and the $FADH_2$ substrates required to keep electron transport going so that thermogenesis can occur.

3. In the skunk cabbage, as temperature decreases, the need for thermogenesis increases. Thus the rate of aerobic oxidation of glucose increases to increase the rate of flux of $NADH/H^+$ and $FADH_2$ through the electron transport chain. Since oxygen is the final electron acceptor in the electron transport chain, an increase in the rate of substrates through electron transport will also increase oxygen consumption. In the lotus flower, when the ambient air temperature is warmer, the need for heat production is less, so there is less flux through the electron transport chain. Thus oxygen consumption decreases.

4. The synthesis of uncoupling protein increases with decreasing temperature, presumably by an increase in transcription of the mRNA that codes for the uncoupling protein. The results shown in Figure 24.1 indicate that the synthesis of mRNA is increased when temperature decreases. The increased amount of mRNA likely results in an increase in concentration of the uncoupling protein. The uncoupling protein would then dissipate the proton gradient which would lead to the thermogenesis observed in plants at cold temperatures.

Amino Acid Role in Glucagon Receptor Binding and Signal Transduction

1.

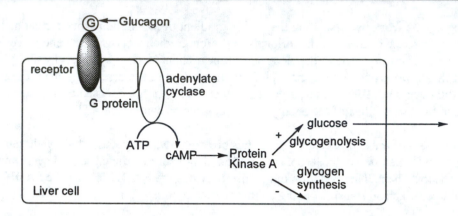

Glucagon binds to the extracellular domain of the receptor in the liver. Binding of glucagon activates the G-protein, which in turn activates adenylate cyclase, which converts intracellular ATP to cAMP, the second messenger. Cyclic AMP activates Protein Kinase A, which phosphorylates several proteins which leads to the stimulation of glycogenolysis and the inhibition of glycogen synthesis. The end product of glycogenolysis, glucose-1-phosphate, is isomerized to glucose-6-phosphate. The phosphate is removed and glucose leaves the liver and enters the blood.

2. These amino acids are positively charged. Since a negatively charged aspartate residue in the glucagon receptor molecule has been shown to be essential for binding, it's possible that an ion pair forms between a positively charged amino acid side chain (His, Lys, Arg) and the essential arginine. This hypothesis can be tested by modifying His^1, Lys^{12}, Arg^{17} and Arg^{18} to neutral or negatively charged side chains and assessing the resulting analogs' binding and signal transducing capabilities.

3. a. Eliminating the Asp at position 9 results in an analog with decreased affinity for the receptor, but little biological activity, indicating that the Asp plays a role in *both* binding and signal transduction. Substituting the Asp with a positively charged Lys decreases the binding affinity by about half, but completely eliminates the biological activity. The Asp evidently plays an important role in binding, but conservation of the negative charge does not seem to be critical since a positive charge does not abolish binding, so some other aspect of the Asp side chain structure is important for binding. The Asp at position 9 does seem to be important in biological activity, since deletion or substitution of the Asp greatly decreases biological activity.

   b. Abolishing the positive charge at position 12 decreases binding affinity by 50-90%. But once the analogs are bound, they are still capable of eliciting a biological response. The more nonpolar analogs (Ala and the acetylated Lys) are able to bind more effectively than the Gly^{12} analog indicating that nonpolar interactions between the hormone and the receptor are important. The addition of a negative charge at position 12 virtually abolishes binding, so it's possible that the positive charge at position 12 forms an ion pair with a negatively charged amino acid on the glucagon receptor.

c. Leu[17] binds more effectively to receptors than does Ala[17], and once bound, has greater activity. This supports the hypothesis from part a. that hydrophobic interactions between the hormone and the receptor are important, since leucine has a more hydrophobic side chain than alanine. Substitution with a Glu residue also decreases binding, but not as much as position 12.

d. These results confirm the hypothesis that hydrophobic interactions are important, since substitution with an alanine leads to a greater decrease in binding than substitution with leucine. The positive charge is important, since replacement with the negatively charged glutamate abolishes more than 90% of the binding ability of the analog.

e. The des-His[1]-glucagon has decreased ability to bind and also has decreased biological activity, but the biological activity is decreased more than binding when the histidine is removed. This indicates that the histidine at position 1 is important both for binding and for signal transduction, but the histidine plays a greater role in signal transduction. This is also supported by the additional des-His[1] analogs. The des-His[1]-des-Asp[9] analog does not bind well (only 7% of the control) and has no biological activity. The binding and biological activity of the des-His[1]-des-Asp[9] analog are lower than the des-Asp[9] analog, indicating that histidine is important in both binding and signal transduction. Interestingly, the des-His[1]-Lys[9] derivative binds well (70%) but has no biological activity. This indicates that the substitution of aspartate for lysine at position 9 has retained whatever characteristics are important for binding that the aspartate also possesses. However, once bound, signal transduction does not occur. The des-His[1] derivatives that include a substitution of a negatively charged Glu for either of the positively charged residues at positions 12, 17, and 18 result in analogs that do not bind to the receptor.

4. The negatively charged Asp at position 9 is important in binding of glucagon to its receptor but appears to play a greater role in signal transduction. This also appears to be true for the His at position 1. Positively charged amino acid residues are important, but not absolutely essential for binding of glucagon to its receptor. It's possible that an ion pair forms between the positively charged amino acid residue and a negative charge on the receptor protein. However, ion pairs are not the only important interaction. Hydrophobic interactions between the hormone and its receptor also play an important role in binding. The positively charged amino acids at positions 12, 17, and 18 are important for binding, but not as important for biological activity. The analogs had difficulty binding to the receptor, but once bound, elicited a full (or nearly full) biological response.

5. The des-His[1]-Lys[9] is the best antagonist because it binds to the receptors with 70% of the affinity of the native hormone, but has no biological activity. In this derivative the two amino acids important in signal transduction have been modified while the positively charged residues at positions 12, 17, and 18 which are critical for binding have been retained. It might be possible to synthesize an even better antagonist (one with greater binding affinity) by trying various amino acid replacements at the 9 position. The replacement amino acid must be chosen carefully so that the characteristics required for binding at this position are retained while the characteristics required for signal transduction are eliminated.

Case 28
The Bacterium *Helicobacter pylori* and Peptic Ulcers

1. Proteins in the bacterium *H. pylori* have a high content of the basic amino acids lysine and arginine. The sequencing of the genome has revealed that proteins in the bacterium contain twice as many of the basic lysine and arginine residues than proteins in other bacteria, perhaps as an adaptation to the acidic environment. As a result, these proteins would have high pI values (pI > 7).

2.

$$NH_2 - \overset{\overset{\displaystyle O}{\|}}{C} - NH_2 \ + \ H_2O \ \xrightarrow{\text{urease}} \ 2\,NH_3 \ + \ CO_2$$

urea

The ammonia is a base and can accept protons to become ammonium ions. By binding protons, the pH in the microenvironment is increased slightly.

3. a. A transporter might import protons or other positively charged ions.
   b. Anion transporters might transport negatively charged ions out of the cell, leaving the interior more positively charged.

4.

Case 29
Pseudovitamin D Deficiency

1. Because the patient's serum levels of 25-hydroxycholecalciferol are high and levels of 1α, 25-dihydroxycholecalciferol are low, the patient is probably suffering from a 1α-hydroxylase deficiency. The patient's serum calcium levels are lower than normal because, despite the fact that the patient has adequate calcium in his diet, the deficiency of the active form of Vitamin D means that the dietary calcium is not being adequately absorbed.

2.

Patient	Base Pair Location	Mutation	Reading frame	Amino Acid Change
Justin	319	G → A	..CCC|GAG|CAC|TGC|AGC...	Arg → His
Patient A	374	G → A	...GCT|TGC|GAA|CTG|CTC	Gly → Glu
Patient B	1004	G → C	...CTC|TCC|CCG|GAC|CCC..	Arg → Pro
Patient C	1144	C → T	...CTG|TAC|TCT|GTG|GT..	Pro → Ser

3. Add radioactively labeled substrate, 25-hydroxycholecalciferol, to samples of cultured control cells and cultured cells expressing the mutated 1α-hydroxylase enzyme. Incubate the cells and then take samples of the culture media at timed intervals and assay for the presence of the expected radioactive product, 1α,25-dihydroxycholecalciferol. The control cells should have high levels of 1α,25-dihydroxycholecalciferol whereas the cells expressing the mutated protein will show no detectable product. These results indicate that the mutated 1α-hydroxylase enzyme is nonfunctional.

4. Justin's enzyme shows an Arg → His replacement. Although both amino acids are basic, arginine carries a full positive charge at physiological pH whereas histidine is only partially protonated. The arginine in the native enzyme may participate in an electrostatic interaction with another amino acid side chain in the enzyme that is crucial to maintaining the proper three-dimensional structure of the enzyme.

The other patients have mutations that are also likely to cause changes in the three-dimensional structure of the enzyme. Patient A has a change of Gly to Glu. Glycine is a small amino acid whereas Glu is much larger and negatively charged. Patient B has an amino acid change from Arg to Pro. The proline will likely change the secondary structure of the polypeptide backbone, since it is a secondary amine rather than a primary amine. Patient C has an amino acid change from Pro to Ser. In the same manner, the Pro in the native enzyme might have been essential for maintaining a certain polypeptide backbone shape.

Case 31
Hyperactive DNAse I Variants–a Treatment for Cystic Fibrosis

1. a. In all the variants, a neutral or negatively charged amino acid has been replaced with a positively charged amino acid (Lys or Arg). The +1 abbreviation means that one additional positive charge was introduced, a +2 indicates the introduction of two additional positive charges, etc.
    b. DNA is negatively charged because of the phosphodiester backbone. The protein engineers reasoned that an enzyme with an increased number of positive charges would bind more effectively to the negatively charged DNA.

2. Intact, double-stranded DNA has a lower absorbance at 260 nm than single-stranded DNA. An increase in absorbance at 260 nm over time would be a useful measurement of the catalytic activity of DNAse I, since the products of the reaction are short, single-stranded oligonucleotides.

3. All of the variants have lower K_M values than the wild-type DNAse I enzyme. This means that the DNAse I variants bind more tightly to the DNA substrate than the wild-type enzyme. The tighter binding is no doubt due to the additional positive charges on the variants which allows the formation of ion pairs between the variant enzymes and the DNA. There seems to be a rough correlation between number of positively charged residues and K_M value–an increase in the number of positively charged amino acid residues results in a lower K_M value, which indicates tighter binding. However, the v_{max} values must also be assessed. All of the variants (with the exception of a few variants with only one amino acid change) have a higher v_{max} value than the wild-type enzyme, indicating that the tighter binding leads to greater catalytic activity. Three amino acid changes seem to optimize K_M and v_{max} values. When four or five amino acids are replaced, the velocity is lower than when only three amino acids are replaced. Since the K_M for the +4 and +5 variants is also generally smaller, it's possible that the enzyme and substrate bind to each other too tightly and that catalytic efficiency is compromised as a result.

4. The pBR322 plasmid DNA normally exists in a supercoiled circle, as shown in the control lane. The wild-type DNAse I can nick the DNA on one backbone to convert the plasmid to the relaxed circular DNA. The +3 mutant has the best ability of the three mutants to carry out hydrolysis reactions on both DNA strands. Cutting both strands results in linear DNA, which is formed in the greatest amount with the +3 mutants than for the other mutants. But all mutants show some ability to produce linear DNA whereas the wild-type DNAse I does not. This indicates that the variants can cut both strands whereas the wild-type enzyme cuts only one strand.

5. All of the mutants have a greater nicking activity than the wild-type for low and high molecular weight DNA at low concentrations. The +2 mutant is especially active under these conditions and would be a good choice to use for a lupus patient. The +1 and +2 mutants can degrade low molecular weight DNA at a high DNA concentration better than the wild-type, but the +3 and +4 mutants are not as active. Not all of the data is given so it is difficult to make conclusions about the ability of the mutants to degrade high molecular weight at high concentrations, but the +2 mutant is clearly more effective than the wild-type DNAse I, and would be a good choice to use for a CF patient.

6. If three amino acids in DNAse I are replaced with positively charged amino acids, the resulting variant is hyperactive–it catalyzes the reaction with a lower K_M and a greater v_{max}. The decreased K_M indicates that the enzyme has a high affinity for its substrate. The increased affinity is due to the replacement of the neutral or negatively charged amino acids with positively charged amino acids that can form ion pairs with the negatively charged phosphodiester groups on the DNA. Increased affinity leads to increase in rate of hydrolysis of phosphodiester bonds, as shown in Figure 21.2. If the enzyme and the substrate bind more tightly the enzyme might be able to make two cuts instead of one. The +3 mutants have the best ability of all of the mutants at making "cuts" rather than nicks. The +4, +5 and +6 mutants clearly have the ability to bind DNA, but their affinity for the substrate is perhaps too great, which causes the v_{max} values to decrease. The +2 mutants have a good ability to nick low molecular weight DNA even at low concentrations.

NOTES

NOTES

NOTES

NOTES

NOTES

NOTES

NOTES